INFORMATION TECHNOLOGY APPLICATIONS IN TRANSPORT

TOPICS IN TRANSPORTATION

Also in this series
Predicting Intercity Freight Flows
Patrick Harker

Books of Related Interest
New Survey Methods in Transport
2nd International Conference,
Hungerford Hill, Australia, 1983

Proceedings of the Ninth International Symposium
on Transportation and Traffic Theory,
Delft, 1984

Behavioural Research for Transport Policy:
The 1985 International Conference on Travel Behaviour,
Noordwijk, 1985

INFORMATION TECHNOLOGY APPLICATIONS IN TRANSPORT

Edited by

Peter Bonsall
Lecturer in Transport Planning,
Institute for Transport Studies, The University of Leeds, UK

and

Michael Bell
Lecturer in Transport Operations, Division of Transport Engineering, Department of Civil Engineering, The University of Newcastle upon Tyne, UK

Utrecht, The Netherlands

VNU Science Press BV
P.O. Box 2093
3500 GB Utrecht
The Netherlands

First published 1987

CIP-DATA KONINKLIJKE BIBLIOTHEEK, DEN HAAG

Information technology applications in transport / ed. by
Peter Bonsall and Michael Bell. — Utrecht : VNU Science
Press. — I11. — (Topics in transportation)
ISBN 90–6764–066–2 bound
SISO 657 UDC 656.011.56
Subject heading: transport ; information technology.

Printed in Great Britain by J. W. Arrowsmith Ltd., Bristol

CONTENTS

PREFACE

It is, of course, impossible to cover in one volume all aspects of the application of Information Technology to transport. Out selection was intended to cover a number of issues from the initial collection of data, through its use in planning and control, to its use in marketing and demand management. Our inclusion of chapters concerned with each of these issues and with a variety of modes of transport will, we hope, enable the specialist or student whose prime interest lies in one area, to draw useful parallels and inspiration from others. By incorporating contributions from specialists from a variety of backgrounds we have ensured that their different perspectives and up-to-date knowledge are fully represented.

The book is aimed at the serious student of transport systems seeking information on techniques used within the industry, and the specialist practitioner from one part of the industry seeking a description of related fields with a view to the development of linked systems, or seeking inspiration from the methods adapted by specialists in other areas. At the same time we believe that the book will appeal to the technically minded layman keen to discover something of the systems that support the smooth operation of transport services.

We will have succeeded in our objective if the reader, drawn to the book by the prospect of one or two of its component chapters, takes the opportunity to dip into others and finds the experience as interesting, instructive and enjoyable as we have found compiling them.

Our most difficult task in compiling the book was not what to include but what to leave out. We have drawn attention to some of the more significant exclusions and to some common themes via the editorial chapter and the editors' notes which follow each of the specialist chapters.

Apart from the editors' notes, our involvement in the individual chapters rarely went beyond setting the original briefs, inviting the authors and commenting on initial drafts. We have not for example attempted to impose a uniform style of writing, nor a uniformity of views on important issues.

Finally we welcome this opportunity to record our thanks to the authors for agreeing to participate in this venture (and to meet some very tight deadlines despite their own very busy schedules). We would like to express our particular admiration for those authors whose mother tongue was not English. We would like to thank colleagues and associates for their encouragement, support and assistance, particular thanks in this respect to Calum Naylor, Bruce Taylor, John Wootton and Michael Heaton. Last, but not least, we could not have met the deadlines without the patient understanding and help of Gillian, Sarah, Marian, Linda, Reiko and Margaret.

The Editors

Information Technology Applications in Transport, pp. 1–10
P. Bonsall, M. Bell (Editors)

Chapter 1

EDITORIAL — ISSUES IN THE APPLICATION OF INFORMATION TECHNOLOGY TO TRANSPORT

Michael Bell[1] and Peter Bonsall[2]
[1]*University of Newcastle, UK;* [2]*University of Leeds, UK*

The phrase 'Information Technology' defies universally acceptable definition but clearly encompasses the methods of collecting, processing and disseminating information as well as its use in system design, management and control. As such it obviously has numerous applications in the field of transport. There is a growing international awareness of the contribution which information technology is making in a variety of industries. This awareness has been promoted by large scale initiatives, for example the European ESPRIT programme and the selection of a transport theme for the 1986 World Exposition in Vancouver, Canada. At the national level the designation of Information Technology Years, or more tangibly, the direction of funds into programmes such as that overseen by the UK Alvey Directorate or such as the development of the Fifth Generation computer in Japan, have done much to stimulate this awareness.

In the field of transport the application of information technology has been stimulated by developments within the industry itself and by encouragement from outside bodies. For example, in the UK, the Science and Engineering Research Council has established a special programme to promote University research on information technology in transport. The programme, which commenced in 1983, has supported a variety of projects concerned with, for example, vehicle and crew scheduling, real-time information systems, passenger advice systems, incident detection, route guidance, data collection technology and expert systems. Both editors of this book are active in the programme as well as more generally in research and teaching related to the application of information technology to transport.

STRUCTURE OF THE BOOK

The book includes specialist contributions covering a variety of areas of application. Broadly speaking the book has been arranged such that the chapters concerned with data collection are followed in turn by those dealing with system monitoring and control, those dealing with optimisation of operations and finally, those dealing with user information systems. We conclude with a chapter on the possible role of expert systems in transport.

In order to draw attention to some of the many linkages between chapters and to the common themes which they address, editorial notes have been provided at the end of each specialist contribution. These notes are also used to provide reference to additional material where appropriate. Parenthetic numbers in the text margins are used to draw attention to these notes.

The remainder of the current chapter is used to highlight some of the themes which we regard as particularly significant.

ISSUES

Man-machine interface

User friendliness is clearly an important consideration in the industry's decision to adopt information technology. In many applications the user's inferface with the system is through a computer terminal. Recent commercial software makes extensive use of facilities such as menu-driven interactive operation, graphic screens and tablets, 'mice' or touch-sensitive screens. The development of automatic voice recognition and synthesis is also well underway.

Several chapters refer to the importance and usefulness of such developments. Chapter 11 shows how colour graphics can assist in the planning of routes and schedules. Chapters 7, 8, 9 and 10, among others, make reference to the potential of graphic displays in the planning and control of operations, and Chapter 15 discusses the importance of user friendliness in the context of expert systems.

Several of the existing applications of information technology to analytical problems deal with simplified, idealised situations because the more complex problems are as yet intractable (see below). In such circumstances it is clearly important that the technician be able to interact with the system in order to modify proposed solutions so as to ensure compliance with those constraints that were deemed too complicated for inclusion in the initial problem. Graphical editing facilities are proving particularly valuable in this respect.

Much of the impetus behind the application of information technology to data collection or monitoring stems from the desire to spare technicians from tedious and repetitive work. There is the obvious risk of an inattentative or overstretched operative failing to register or record important information or events. Chapters 2, 3, 4 and 5 are all concerned with systems to record data with minimum human involvement. In complex real time applications, such as piloting aircraft, there is an increasing emphasis on systems which take over the routine monitoring role leaving the human (in this case the pilot) to concentrate on decision-making in the light of data brought to his attention.

It is accepted that automatic pilot systems can deal quite adequately with routine situations, but it is felt that no fully automatic system can yet be trusted to deal with unusual or complex situations such as emergencies where the costs of a wrong decision may clearly be high. The possible future role of expert systems in this context is discussed in Chapter 15 but, for the foreseeable future, automatic systems are unlikely to be adopted. In contrast to this, arguments are put forward in Chapter 6 to suggest that, in the case of motorway surveillance and control, a fully automated system is to be preferred, because speed of reaction to an incident is of critical importance and there is every reason to believe that the system could perform as reliably as a human operator given the same information.

An important consideration in the man-machine interface is the amount of information that is fed back from the machine to the user. In the context of expert systems, it has been argued that the reasoning behind conclusions or recommendations should be shown, to facilitate checking and credibility. However, the converse has been argued in the context of traffic control (see Chapter 6). When motorway traffic is guided by variable message signs, it has been suggested that the driver should be set a maximum speed with no justification offered, for the reason that if drivers are informed about the conditions ahead, they may select divergent speeds, possibly resulting in a more dangerous situation.

User friendliness is particularly important where there is a direct interface between the system and the public. See, for example, Chapters 5, 13 and 14 which variously deal with ticketing systems, user information and route guidance. Negative public reactions to such systems are sometimes attributed to a phenomenon depreciatingly known as 'techo-fear' (an irrational fear of technology). In practice, of course, the negative feelings may be a rational response to systems whose performance or basic design have yet to instil confidence.

The introduction of voice synthesis for systems which involve disseminating information to members of the public is seen by some as particularly valuable in 'humanising' the machines. Others, however, see a risk of an opposite reaction.

Centralisation versus decentralisation

Early examples of the application of information technology to transport were characterised by centralised processing, while more recent systems have relied to a greater extent on distributed processing. The move from the centralised to the distributed systems is illustrated in several chapters. The reasons behind this trend include the fact that as the centralised systems grew in complexity, the load on communication channels between the central computer and the numerous users or outstations became a significant limitation. Furthermore, centralised systems, particularly in the absence of backup facilities, can be completely paralysed by a failure at the centre. The advent of microprocessors in particular allowed some of the load to be taken off the central computer and the communication links.

An interesting example of this trend is exhibited by urban traffic control. In on-line systems for controlling a network of signals, such as SCOOT (described in Chapter 6), processing is located in a central computer and dedicated lines of communication to outstations are required. Interest is, however, currently turning to enhanced Outstation Monitoring Units (OMUs) linked to a control centre via the telephone network. Although OMUs were originally designed for fault monitoring, whereby the centre is autodialled in the event of a fault, more recent versions can also be used to initiate signal plan changes. Thus, a degree of centralised coordination is achievable when necessary, while leaving each installation otherwise independent to adjust itself in the light of local traffic conditions.

Institutional Issues

It is clear that the implementation of information technology is subject to substantial institutional pressures and constraints. For example, the trend towards distributed computing facilities, referred to above, has been given extra impetus in some industries by the desire of local managers to circumvent the central rigidities and inertia which had restricted the access to computing facilities controlled by monolithic data processing departments (see, for example, Chapter 10 in respect of such problems in the bus industry). Where the traditional management structure has devolved considerable decision-making to the local level, the introduction of distributed micro computer systems is clearly less disruptive than would be an attempt to introduce centralised processing (Chapter 7 deals with such considerations in relation to British Rail). The ability of local managers to use their micros for a multiplicity of purposes is clearly important and is specifically mentioned in Chapters 5, 7, 10 and 12.

The reaction of workers and managers to the introduction of information technology in their industry is often coloured by the prospect of inevitable changes in work practices and the redundancy or redefinition of some traditional expertise. Increasingly, the skills at risk include not just those of a routine nature but, particularly with the advent of expert systems, those involving substantial training and experience.

There are several examples of information technology systems which, although technically feasible, cannot be justified economically. More interestingly there are several systems where the benefits are demonstrable but where problems of cost allocation or revenue collection have yet to be overcome. Examples include systems where the benefits would be available to all-comers whether or not they had paid for them (e.g. broadcast traffic information). In such circumstances market forces may fail to ensure development and substantial public investment may be required.

Optimisation

Many examples of the application of information technology to transport involve an attempt to optimise in complex situations. However, because of the complexity, the problem is often reduced to a sequence of tasks which are optimised independently. In the presence of feedback between the tasks (that is, where the outcome of a later task influences the optimum outcome of an earlier task), piecemeal optimisation may fail to yield a global maximum.

Nevertheless, there are several industries in which satisficing behaviour still predominates. That is, one starts with a feasible solution and occasionally makes ad hoc improvements to it, in response to particular bottlenecks or constraints which arise as circumstances change. This may be the best currently available strategy for certain very complex problems.

Elsewhere, for example in freight routeing as described in Chapter 9, heuristics (simple rules which drastically reduce the size of the problem and yield a good, but not necessarily optimum, solution) have proved very valuable. Increases in computing power are likely to lead to the replacement of heuristics by strict optimisation involving combinatorial (or integer programming) methods. Indeed for some problems (see, for example, Chapter 10) such approaches may already be feasible.

Interaction with demand

An interesting question is whether the application of information technology to transport will stimulate or replace physical travel. Clearly many applications described in this book (for example, route guidance, passenger information and traffic control systems) may make travel more attractive and, therefore, might stimulate demand. The development of information systems in the travel industry (see Chapter 12) is clearly based on this assumption. Other applications, however, could obviate the need for certain types of trip. The net effect is hard to assess.

Much has been written on the potential impact of increasingly sophisticated telecommunications services on the demand for travel. Early analyses suggested that a substantial amount of business travel might be replaced by tele-conferencing, that numerous workers might keep in touch with their city centre workplaces via networked micro computers, thus obviating the daily journey to work, and that shopping trips might be replaced by armchair shopping via interactive viewdata services.

More recent research based on experience to date with such services has suggested that, although some substitution of telecommunication for physical travel can be expected, the new media will more often serve a previously unmet need with very little reduction in physical travel. Thus tele-conferencing allows extra staff to be involved in meetings at a low marginal cost and, rather than substituting for such meetings, facilitates brief preliminary meetings prior to more substantial ones.

More generally, history suggests that telecommunications and travel are synergistic rather than substitutable; improved telephone networks have often been followed by increased travel and trade, while improved physical communications (e.g. the road bridge between England and Wales across the Severn Estuary) have been followed by increased telephonic traffic. The relationship is clearly very complex. We may expect improved communications to foster closer links between spatially separated areas and the greater availability of services, such as in-vehicle communications using cellular mobile radio, to increase the productiveness of journey time, thus increasing the propensity to travel. Conversely, we must expect some shift from physical commuting to telecommuting.

It is a brave man who is prepared to predict the net impact of such developments or to speculate on whether, if teleshopping becomes important, it will do more than substitute for existing mail order shopping and, if it does, whether retail delivery patterns will be altered radically. In fact, of course, all these changes are likely to be associated with, and confused by, changes in the economic and social climate.

Communications media

We have referred, in a previous section, to the importance of a good man-machine interface and to the value of interactive media. Several of the chapters make reference to various forms of videotext system and we have thought it appropriate to list here some of the more important systems and to offer some definitions in what is sometimes a rather confusing field.

VIDEOTEXT is a global term covering two types of on-screen information system; TELETEXT and VIDEOTEX. The TELETEXT systems are broadcast alongwith television signals. One well-known example is

CEEFAX, which is broadcast in the UK by the BBC, and another is TEXT-TV, which is broadcast in Sweden. TELETEXT systems have very limited capacity, have slow response and are not truly interactive in that the user cannot send information back to the system.

VIDEOTEX (sometimes known as VIEWDATA) systems, on the other hand, are truly interactive. They provide access via cable, telephone or satelite links to data bases and processing power resident on host computers. The user can interact with the system to seek information, to make transactions or to make use of the computing facilities available. Links from one part of a system to another are provided via 'gateways'. Some VIDEOTEX systems are public while others are restricted. The most important public system in the UK is PRESTEL, operated by British Telecom. Examples of similar systems elsewhere include VIDITEL (Netherlands), CAPTAIN (Japan), BILDSCHIRMTEXT (Germany), MINITEL (France), VIEWTRON, THE SOURCE and COMPUSERVE (USA).

Note that the term TELETEX refers to an international system of communications protocols rather than a VIDEOTEXT system.

CONCLUSIONS

Having identified various themes in the preceding sections, those that we feel to be particularly significant for the future development of transport systems are reiterated here. The first is the trend towards greater decentralisation in processing power. This brings with it the general problem of machine-machine communication, networks, protocols, international standards, and machine compatibilities.

A second area where we see considerable scope for development is the application of statistical techniques for real time system control. A technique of particular significance here is Kalman filtering which, after successful applications in a number of areas including the monitoring of traffic in tunnels and incident detection, is now being considered for a range of other applications including the prediction of vehicle arrival times in the context of a bus passenger information system.

A third area where we anticipate substantial progress is that of coordinated monitoring and control systems providing on-line information to management and users.

A fourth area of considerable significance is that of expert systems. We expect that, after a period of experimentation involving false starts and blind alleys, the potential role of expert systems in the transport industry will begin to be realised. The most useful developments in this area in the short term are likely to be those that are unspectacular but sound.

Finally, we should reiterate our belief that the pace and direction of developments will continue to be determined by economic and institutional pressures, rather than by technological advance.

Clearly any book on the subject of information technology has a degree of built-in obsolescence. Nonetheless, this and subsequent chapters present many themes and concepts that will endure and that are important to an understanding of current developments in the transport industry.

Information Technology Applications in Transport, pp. 11–40
P. Bonsall, M. Bell (Editors)

Chapter 2

VEHICLE DETECTION AND CLASSIFICATION

Peter Davies
University of Nottingham, UK

Recent developments in microprocessor and traffic sensor technology have for the first time made several forms of automatic vehicle classification possible for traffic control and monitoring. This chapter reviews both established techniques and new approaches to vehicle detection, and illustrates how these techniques are being applied within real-time classification systems.

Until recently, manual counts have provided the only source of classified traffic flow figures, while speed distribution data have been based upon radar or other semi-automated measurements. Axle loads have rarely been surveyed systematically; what information there is has often been gathered from inherently biased and labour-intensive static weighing at the roadside. Finally, turning traffic flows at intersections have been based on short-period manual counts as highway authorities responded to specific requests for data.

Current demands for classified traffic count data are already diverse. The long-term monitoring of trends by type of vehicle, for example, provides a basis for statistical forecasts of future vehicle-kilometres. More localised classified counts along highway links or at intersections are commonly required for highway scheme appraisal or traffic management purposes. Speed classification may be undertaken as a part of the assessment or revision of speed limits, or to monitor the effectiveness of enforcement measures. Weight data may similarly be required for enforcement appraisal or for the determination of standard axle equivalent traffic flows used in highway pavement design. To satisfy these increasing demands for data, highway authorities are developing structured systems of traffic monitoring, utilising new techniques founded on information technology (Gruen, 1985).

Conventional traffic survey methods are expensive. Traffic patterns can be highly variable, resulting in a need for careful survey design. While new techniques permit different types of data to be collected more reliably and at lower cost than before, questions of sample design which are outside the scope of this chapter must still be addressed. Allsop (1984), Taylor (1984), and Richardson and Taylor (1984) provide recent papers setting out a systematic approach to survey design, needs appraisal and sample selection through time and space. This review concentrates on survey methods and techniques for vehicle detection and classification.

The chapter begins with an examination of conventional pneumatic tube and inductive loop vehicle detection techniques. It goes on to examine alternative sensors for vehicle detection, before looking at the development of automatic classification systems. Current systems for vehicle classification by lane, speed and type are examined, as well as the scope for further developments.

VEHICLE DETECTION

Vehicle detection is the ability to sense and indicate the presence or passage of a vehicle. The first vehicle detectors were developed for vehicle actuation at traffic signals, for which pneumatic systems became widely used in some countries. Pneumatic detectors were later adapted for traffic counting using portable tubes and electro-mechanical counters. More recently, classification of vehicles by speed, weight or type has become a practicable proposition.

Distinction can be usefully drawn between sensors, detectors and monitoring systems. The sensor can be defined as a transducer placed in, on, or near the roadway in order to sense the presence or passage of a vehicle. The detector can be defined as electronic circuitry that is connected to the sensor in order to translate the sensor output into an indication of the vehicle. The monitoring system is the incorporation of sensor, detector and a medium for data analysis and storage or transmission, enabling use to be made of the vehicle detection.

Technological advances in all three areas have been obvious over the past thirty years. Improved sensors have allowed more precise vehicle discrimination. Improved detectors have enabled a wider range of sensors to be used for a number of different purposes, and improved monitoring systems have allowed data from any given sensors and detectors to be collected and utilised more easily and reliably.

The first vehicle detection techniques were solely manual. They took the form of manual counts recorded on paper for subsequent analysis. Mechanical aids such as tally counters were subsequently developed to reduce fatigue and hopefully increase the count accuracy. The deficiencies in manually observed data are a function of the problems associated with supervising a large and generally uninterested field force, coupled with the need to restrict the counts to short periods for reasons of cost and practicality. The UK Traffic Appraisal Manual (Department of Transport, 1981) shows that manual enumeration by class may give rise to standard errors as high as $\pm$ 28% for certain vehicle types.

Pneumatic Tubes

The need to replace manual with automatic sensing became apparent in the 1920s when traffic signals were replacing manually controlled intersections. The first commercially successful vehicle detection system came in the form of the electro-pneumatic sensor developed in the 1930s. This device consisted of thick-walled rubber tubes laid in a channel in the pavement, terminated by a diaphragm switch detector. The passage of an axle over the sensor generated a pulse of air in the tube which was detected by the diaphragm switch. The system gained wide acceptance in Britain and elsewhere for vehicle actuation of traffic signals. Its inability to detect vehicles that are stationary over the sensor, its high cost of installation and its susceptibility to damage led, however, to a decline in the use of this technique as alternatives became available.

Despite this decline in use for permanent detection, the electro-pneumatic sensor was successfully transposed to the field of temporary vehicle detection in the form of the pneumatic tube. This "O" or "D" section tube has been used extensively in road traffic data collection. Griffiths (1982) described early units which were simply counters recording total volume only, measured in axle-pairs. Later adaptations such as the Fischer and Porter counter, described by Blackmore (1966), incorporated clock mechanisms and a recording medium, usually paper tape, for the storage of totals by time period.

Pneumatic tube counting technology has improved considerably since the days of electromechanical diaphragm switches. Problems of switch failure and under or overcounting have been reduced by the adoption of solid-state piezo-electric pressure detectors. A small disc coated with a piezo-electric ceramic produces an electrical charge when deflected by an air pulse. The charge is processed by transistor circuitry into a square-wave signal of 20 to 30 ms duration. The rapid response and variable dead time reduce the chances of undercounting closely-spaced fast axles or overcounting echoes in the pneumatic tube, while improved sensitivity means that vehicles can be counted down to walking pace or less.

Pneumatic tube-based systems have, for many years, been considered the most suitable for temporary vehicle detection and counting. In the UK, for example, they have been employed in the collection of data for the 50 point continuous census for more than twenty-five years. They have also been used extensively by highway authorities for local data collection and traffic monitoring. Their cost is low and installation is straightforward, while removal and re-use at other sites presents few problems. Their ability to discriminate between lanes is limited, however, which tends to restrict them to providing nothing more than a total axle count.

Inductive Loops

Duley (1981) describes the inductive loop detector and how it was originally designed for the actuation of barriers in car parks. The loop sensor, whose inductance changes when a vehicle enters its zone of detection is connected to detection circuitry which must respond to that change of inductance but ignore other changes in the loop impedance due to environmental drift.

The first loop detector, manufactured by the Sarasota Engineering Company, was an extension of a textbook inductance bridge circuit. The detection characteristics of this system were unstable and more complex detection circuits have subsequently been developed. Four basic principles of loop detector operation have been discussed by Duley (1981). They all utilise the loop as part of an oscillating circuit whose frequency is monitored. The most recent loop detectors use a digital system where the period for a preset number of loop oscillations is timed by a high-frequency oscillator. The digital loop detectors adopted for continuous automatic vehicle classification at the 240 sites of the UK national core census are examples of this type of system.

For permanent installations, the inductive loop has several distinct advantages over earlier pneumatic sensors. It is relatively easy to install and requires less pavement excavation than the electro-pneumatic sensor. It is not susceptible to damage from vehicles and, if the initial installation is satisfactory, its operational lifetime is long (Head, 1982). Improved electronics with self-tuning capabilities have made systems more robust and less dependent on initial setting procedures. These advantages have led to the replacement of electro-pneumatic sensors with loops for most traffic control purposes.

Temporary inductive loop sensors have been used for short-term counting and classification, by sticking the wires to the road surface. Adhesive loops will stick to a dry, clean road surface, but moisture from the lower layers of the pavement, or creeping under the adhesive by capillary action, can cause loss of adhesion after a few days. Thermoplastic loops, in which the wires are covered by black road-marking material, are tough and longer-lasting but relatively slow to put down. Where speed is important, a better arrangement is to use prefabricated bituthene loops, which can be made up beforehand and put down very quickly provided that the road is dry and free from salt. Their workable life normally ranges from a few weeks through to several months and is most likely to be terminated by damage from stone chippings (Davies, 1983; Kember-Smith, 1985).

Inductive loops operate as follows. When an alternating current is passed through a loop, a fluctuating magnetic field is generated around its wires, extending above and below the road surface. At any instant these "lines of force" may be considered to be similar to those around a bar magnet (Figure 1).

If a metal object enters the field, eddy currents are induced in it, giving rise to their own magnetic fields. These couple with the loop field and reduce the loop's inductance. The amount of this reduction is measured by the loop detector circuitry (Duley, 1981; Moore, 1981).

The size of the inductance change depends upon the strength of the magnetic field cutting the metal object. For a two-dimensional body such as a thin steel plate, the important factor is the component of the field cutting it at right-angles. If vehicles can be considered to consist largely of horizontal and vertical steel panels, the magnetic field can be conveniently treated as three components acting at right-angles to one another. These components act vertically, horizontally along the traffic lane, and horizontally across the traffic lane.

For example, the inductance change produced by a vehicle's horizontal floor-pan would depend mainly upon the vertical component of the loop's magnetic field. The horizontal field along the traffic lane would cut the vehicle's front and back

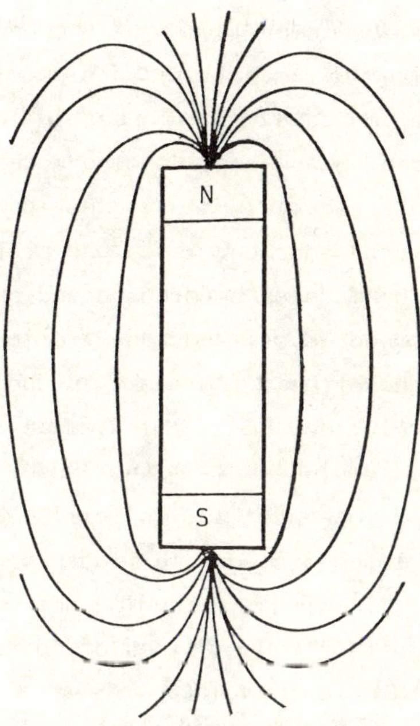

Figure 1 Lines of force around a simple bar magnet

panels, thereby initiating and terminating detection. Bicycles, whose frame and wheels act like a vertical sheet of metal, are detected solely by the horizontal field across the traffic lane. The total inductance change produced by a vehicle is equal to the sum of the component parts.

In their usual applications, inductive loops would ideally possess the following characteristics:

(a) all vehicles, including motorcycles and (perhaps) bicycles, should be detected;

(b) vehicles should not be detected by loops in adjacent lanes;

(c) vehicles should be detected continuously during their passage over the loop;

(d) the points of first and last detection should be independent of the vehicle type and its lateral position; and

(e) the strength of the loop signal should be independent of the vehicle's lateral position.

Broadly, (a) and (b) are required for vehicle actuation, (c) for automatic barrier operation, and all three for traffic counting. Additionally, (d) is of importance in vehicle length measurement and (e) in chassis height or magnetic signature determination. All five characteristics are desirable for automatic vehicle classification.

Existing loop designs for these different purposes have evolved largely as a result of practical experience. The performance of existing loops fails to satisfy any of these ideal requirements fully. With this in mind, some experimental investigations were carried out at the University of Nottingham into the fundamental properties of inductive loop layouts. These laboratory and field tests aimed to see if new loop designs could approximate more closely some or all of the ideal characteristics. A particular aim was to develop an improved loop for vehicle classification.

Full details of the experimental procedures adopted are given by Davies et al (1982). The first series of laboratory tests aimed to establish how the maximum height of the detection zone varies with loop dimensions. A high zone of detection is necessary if high-backed articulated vehicles are to be detected as a single entity, for traffic counting or vehicle length measurement.

The results indicated that, for rectangular loops of a given perimeter, field height is a maximum for maximum enclosed area, i.e. for square loops. Ordinary rectangular loops performed almost as well, with only the long narrow shapes showing a serious reduction in detection height. Increasing the loop perimeter tended to increase the height of detection, but by a less than proportionate amount; a law of diminishing returns appeared to be operating. Increasing the number of loop turns had little effect.

The second series of laboratory tests was concerned with magnetic field contours for different loop shapes. Contours for

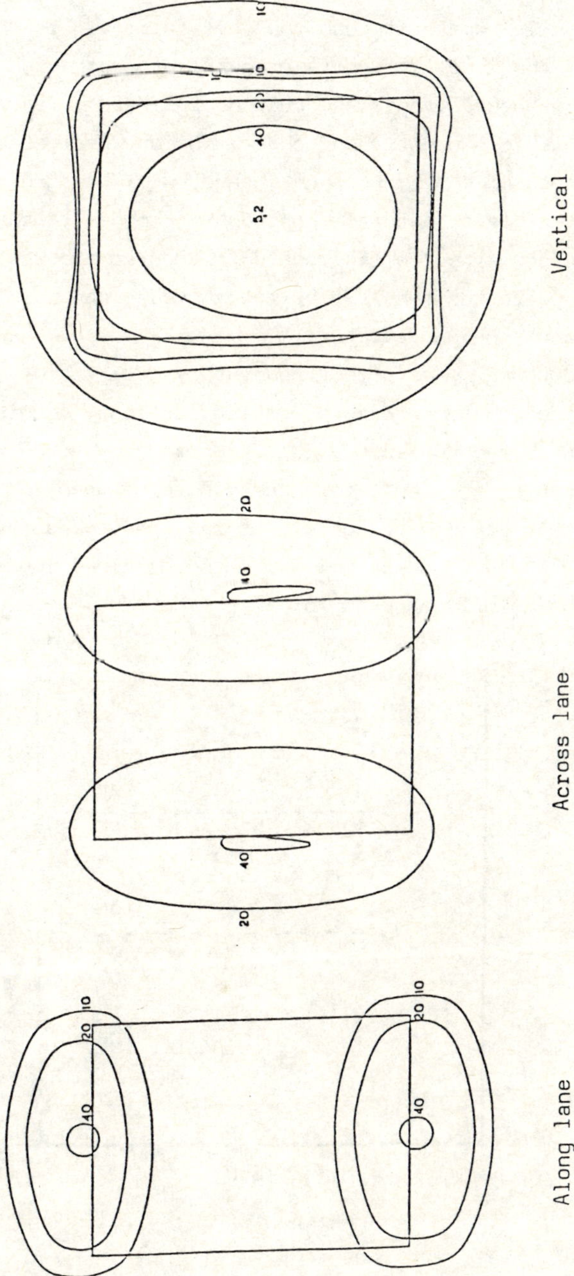

Figure 2 Rectangular loop (three turns)

the rectangular loop in Figure 2 reveal its two main weaknesses. The first is that although the vertical field is high in the centre of the loop, it tapers gradually to each side. Because of this, high-backed vehicles cutting the edge of the loop can produce multiple counts, and chassis height estimates, based on the strength of the inductance change, will be greatly affected by the vehicle's lateral position. Secondly, the horizontal fields across the traffic lane do not cover the whole width of the loop, with the result that motor cycles can pass through its centre without being detected.

The outcome of further laboratory and field tests was a preferred design for a double rectangular loop, consisting of two series-wound rectangles per lane separated by a 200 mm gap. The double rectangle is able to discriminate between vehicle chassis heights over a wide range of lateral positions, as shown in Figure 3. The design is similar to the double-D inductive loop layout subsequently adopted for the UK national core census.

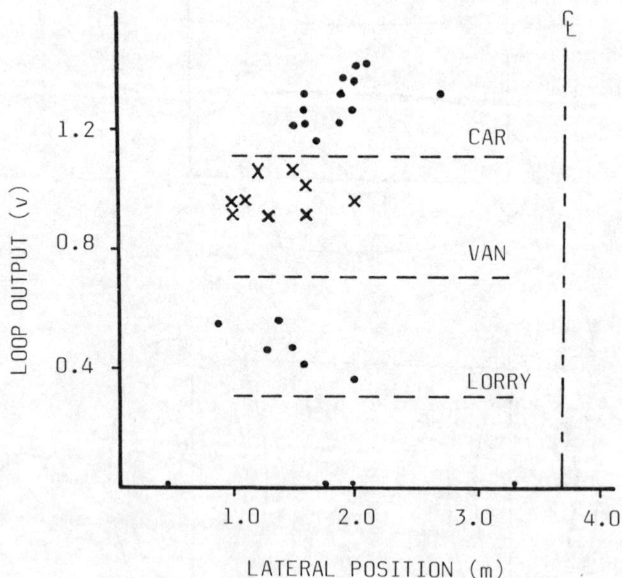

Figure 3 Double rectangular loop used for chassis height detection

ALTERNATIVE APPROACHES

New developments in sensor technology have led to several transducers and other devices that may be regarded as alternatives to the pneumatic tube and inductive loop. Although none has yet been as widely accepted as tubes or loops, a number has been applied with some degree of success. Several of these technologies are specifically applicable to automatic vehicle classification, and are described below. This section provides a summary of some of the current alternatives.

Vehicle Sensors

Optical sensors may take many forms ranging from simple light beams broken by the vehicle passage to complex data acquisition systems using cameras to record vehicle manoeuvres. Honey (1975) has described the use of light beams which are projected across the roadway and either received at the other side or reflected back. Infra-red sensors have been put to similar use at the University of Nottingham. These sensors are relatively simple to install with little disruption to traffic but are susceptible to alignment problems and poor weather conditions, and offer little or no discrimination by lane. Detection in individual lanes may be achieved by projecting the beam vertically downwards but this requires the use of overhead support structures. (1)

Recording vehicles on film or video has proved popular in many research applications. A major limitation of this technique is the level of manual data abstraction and analysis required. Even if a computer is used to assist in the analysis, as described by Wootton and Potter (1981), a significant amount of manual data transfer is required. Dickinson et al (1982) have investigated (2) systems for the automatic abstraction of data from video images, and similar work has been carried out at the Australian Road Research Board (ARRB), as well as under contract to the US Federal Highway Administration (FHWA), but commercial systems are as yet unavailable.

Magnetic sensors that operate by detecting the residual magnetism induced in vehicles by their passage through the earth's magnetic field have been developed by Scarzello et al (1978). An alternative design which operates by generating a magnetic field and detecting the resultant fields produced by eddy currents in the vehicle has been reported by Mills (1974). These types of sensor have not gained wide acceptance but may prove to have distinct advantages over inductive loops for certain applications, having very tightly-defined zones of detection. They are, however, likely to be more costly than inductive loops.

Microwave and sonic detectors operate on the Doppler principle by directing a beam of energy at the road and sensing the energy reflected back by moving objects. Head (1982) indicates that microwave detectors are capable of providing adequate vehicle detection for certain applications. Microwave vehicle detectors (MVD) are widely used in Britain for vehicle actuation of temporary traffic signals and at certain signalised pedestrian crossings. The problems associated with their use have been described by Dickinson (1983). These include the inability of microwave detectors to sense stationary vehicles and the shadowing of smaller vehicles by the presence of a large vehicle in the zone of detection. Current work at the University of Nottingham is examining their use as a stopped vehicle detector within urban vehicle classification systems.

Daniels (1982) has suggested that the position of stationary vehicles and the speed, direction and range of moving vehicles could be determined using continuous wave F.M. microwave sensors, but that the problem of vehicle shadowing is likely to remain a serious drawback.

Axle Sensors

Axle detection requires different approaches from those used in overall vehicle detection. Recent developments have provided several alternatives to the pneumatic tube for temporary and permanent axle detection. Four options based on contact, capacitive, triboelectric and piezoelectric approaches are described below.

Axle weighing presents more specific challenges, including uniformity of response along the sensor length, and independence from temperature, speed and vehicle dynamics effects. Conventional dynamic scales involve significant pavement excavation, with large steel plates supported on load cells or instrumented with strain gauges to give a direct readout of axle load. Two of the sensors described below are actively being developed for low-cost axle load measurement. Further details are presented by Davies and Sommerville (1985).

Contact sensors use the mass of the axle to close metallic contacts in or upon the pavement. The most common form, known as the tapeswitch, is a reusable strip fixed to the road surface, described by Byrne et al (1976). It has been used for a variety of research applications, particularly in the USA, and has proved to be a viable method for temporary axle detection.

Capacitive sensors use dynamic tyre forces to compress two or three parallel conducting plates, thereby changing their capacitance. In a simple form the sensor may be a length of coaxial cable whose inner and outer conductors form the capacitor plates. This system is closely related to the triboelectric sensor described below, which uses different signal processing electronics to achieve a similar end.

A more complex capacitive sensor is described by Basson et al (1977) and takes the form of a portable steel and rubber mat fixed to the road surface. The movement of steel plates under a wheel indicates not only the passage of the axle but also its weight. This type of device has also been considered for permanent axle counting and weighing, through being buried in the pavement.

A recent development from the capacitive mat is a capacitive strip sensor described by Davies et al (1985). Capacitive strips are more portable than mats, and can be manufactured at lower cost. Both counting and weighing outputs can be achieved through the use of a modified signal processing technique using microprocessor equipment.

Capacitive sensors require special signal conditioning systems for axle weighing. The conventional approach is to use the sensor as one element of a tuned circuit whose frequency of oscillation varies with capacitance. Frequency to voltage conversion, with complex feedback and stabilisation circuitry, completes the system by producing an analogue voltage which varies with axle weight. An alternative approach developed more recently is similar to the digital loop detector, avoiding the need for A-D conversion by remaining digital throughout. All tracking and stabilisation functions with the capacitive mat or strip can thus be carried out in software.

The third type of sensor, which is triboelectric, utilises the effect of spontaneous generation of charge with friction between certain materials. These sensors are composed of coaxial cable which exhibit the tribo effect when vibrated or flexed. They may be stretched across the roadway without any mounting for short-term applications, or can be mounted in flexible polyurethane to form a more permanent and robust sensor, fixed into a slot in the pavement surface. Signal processing circuitry for these sensors has been developed at the UK Transport and Road Research Laboratory and, like the detailed sensor design, is covered by patent.

The final approach is the piezoelectric transducer which operates on the generation of charge when certain crystal structures are subjected to stress. The piezoelectric airswitch has been described above, in connection with pneumatic tubes. An alternative form is Vibracoax cable. This commercially available product is a coaxial element, three millimetres in diameter, whose insulator is a piezoelectric powder. The sensor may be surface mounted, or mounted in a slot in the road surface. With suitable charge amplification and signal processing, it can be used successfully as an axle detector.

Piezoelectric sensors are also being adapted for axle weighing. A four year research programme has been carried out at the University of Nottingham under contract to the UK Transport and Road Research Laboratory (TRRL) on the use of this sensor for dynamic axle load measurement, by means of charge amplification followed by fast A-D conversion of signals. Several projects in the United States are currently attempting to take this work to commercial application.

In summary, triboelectric and piezoelectric sensors are becoming quite widely used and are possibly the most suitable approaches for permanent axle detection. Together with capacitive sensors, piezoelectric cable also offers the possibility of low-cost axle weight detection. Inductive loops provide, at present, the most popular solution for vehicle presence detection.

(3)

VEHICLE CLASSIFICATION

In recent years, increasing needs have been felt for the development of comprehensive data collection systems, capable of giving much more information on traffic than simple vehicle flows. Both in Europe and North America, highway authorities have promoted the development of more sophisticated traffic monitoring systems capable of responding to a wide range of specific data requirements. A typical set of needs was summarised by Hillier et al (1978).

Firstly, there is a need for classification of traffic counts by vehicle type. This not only increases the usefulness of information on vehicle characteristics such as speed and headways, but also permits the allocation of construction and maintenance costs to the various classes of road user. Secondly, the tendency towards the use of larger and heavier goods vehicles has increased attention paid to environmental standards and has emphasised the need for data on axle weights. Thirdly, road safety research requires data on volumes, speed and headways, together with associated road configurations, to be extended to cover a wider range of locations and times.

As data requirements become more diverse, the need for

improvements in accuracy and consistency necessitate larger and more representative samples. This reflects the need to improve the prediction of future travel demands by all classes of road user. These requirements for a wider range of traffic data with better spatial and temporal coverage led to the development of several portable or semiportable data collection systems capable of providing such information as speeds, headways and vehicle type.

Early Systems

Van Helden and Van der Voort (1974) describe equipment developed in the Netherlands by the Dienst Verkeerskunde (DVK), the traffic and transportation division of the Dutch Ministry of Transport. The equipment was a mobile data acquisition system housed in a vehicle. The sensor array consisted of two inductive loops per lane which enabled the following parameters to be determined:

(a) vehicle counts per time cycle;
(b) individual vehicle lengths;
(c) individual vehicle speeds;
(d) gaps between vehicles; and
(e) vehicle type (determined from the vehicle length).

Events were timed using loop detectors and subsequently stored on magnetic tape, allowing the required vehicle parameters to be calculated at a later date.

Contemporary work undertaken at the Bundesanstalt fuer Strassenwesen (BAST) of West Germany into the development of an automatic vehicle weighing and classification system was described by Lenz and Hotop (1974). A sensor array comprising inductive loops and a dynamic axle weighing system was used to provide information on vehicle lengths and weights.

Reijmers (1980) reported the development of an on-line vehicle classification system utilising two inductive loops. The analogue voltages from two detector units were processed by a PDP/11 computer to give information on vehicle speed, length and class.

Classification utilised five categories according to a shape factor calculated from the characteristic loop signature for each vehicle. A similar system implemented on a microcomputer has more recently been developed in West Yorkshire, England, for selective vehicle detection within an urban traffic control system, and is the subject of a UK patent application.

The West Yorkshire system acts as a passive bus detector giving priority to public transport vehicles at traffic signals. Digital values of loop inductance are normalised for vehicle speed, and level of loop output, corresponding to different vehicle lateral positions on the loop. Several tests are then applied to identify bus profiles, including number of peaks, edge gradients and the presence of a central plateau within the loop signal. The prototype system overestimated the number of buses by about 14% (Franklin, 1984).

The first comprehensive automatic vehicle classification system was developed at the UK Department of Transport Traffic Control and Communications Division (TCC), and is described by Nash (1976). ALICE (Automatic Length Indication and Classification Equipment) utilised a sensor layout consisting of two inductive loops followed by one axle detector. Events were timed and vehicle parameters calculated by means of a crystal oscillator generating clock signals, registered in various discrete logic counters. The parameters of speed, length, headway, number of axles and their spacing were then printed on a 16 column drum printer to provide a permanent record. The incorporation of logic circuits for vehicle identification allowed real time identification of vehicles into fifteen categories based on axle configuration and overall length.

The digital operations needed for parameter analysis were cumbersome to implement using discrete logic elements and as a result the system had several limitations. Notable among these were its inabilities to register closely following vehicles (less than five metres gap), its inability to register speeds for vehicles travelling less than 13 km/h due to counter overload, and its slow operational speed. It was, however, a major step in the development of a real-time vehicle classification system. The rapid development in microprocessor technology during the late 1970s provided a way forward for the refinement of data collection systems. A microprocessor can accept simultaneous inputs from a number of external sensors, carry out computations using these inputs and produce data output in various forms under user control. Microprocessor traffic monitoring systems offer several advantages over earlier approaches.

Firstly, the use of a sealed unit without any moving parts improves reliability and leads to more consistent results than were obtained from electromechanical or cassette-based recording systems. Secondly, the greater processing power or 'intelligence' available within the microprocessor unit allows preliminary analysis and compression of data as they are collected. The processor can also be used for more complex forms of data collection or classification than had previously been possible, programmed to suit the particular needs of individual users (see, for example, Storr et al, 1979). Finally, the data are created and stored within a computer environment which lends itself to the adoption of further computer analysis techniques, involving a sensible minimum of manual interaction and editing.

Lane, speed and length classification

The simplest form of automatic classification is by traffic lane, using separate loops or tubes to cover each traffic stream. Some less obvious pneumatic tube configurations are also practicable using microprocessor logic in conjunction with a pair of tubes spaced a short distance apart. In Figure 4, for example, directional counts can be obtained even on a single-carriageway road where overtaking is common, using the microprocessor to resolve small time differences between the pneumatic tube signals. Alternatively, if lane counts are required rather than directional flows, the arrangement shown in Figure 5 can be adopted

More complex configurations are used with inductive loops

involving directional logic with pairs of loops in one or more lanes. Microprocessor counters also provide for the use of loop pairs side-by-side in certain lanes at sites where lane discipline is poor, in a configuration commonly referred to as an N+1 layout. When N+1 is combined with directional logic, the number of possible permutations of loop configurations becomes quite large. Some of the possible configurations are described by Davies (1983) and Moore (1984).

Speed classification systems operate by timing vehicles between pairs of sensors spaced up to a few metres apart. Vehicles are counted into separate categories corresponding to different speed ranges. Either pneumatic tubes or inductive loops have been used as sensors; however, loops have the advantages that individual lanes are more easily monitored, and that vehicle, rather than axle, speed distributions are obtained.

Where tubes are used for speed monitoring they must be matched in length and tautness. This is because tube pulses travel at the

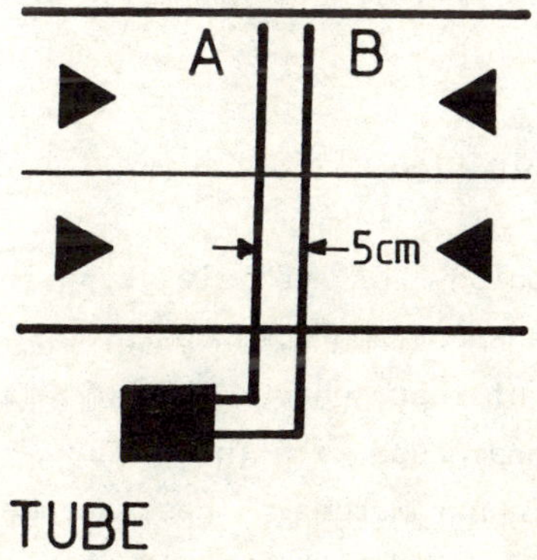

Figure 4 Directional Counting

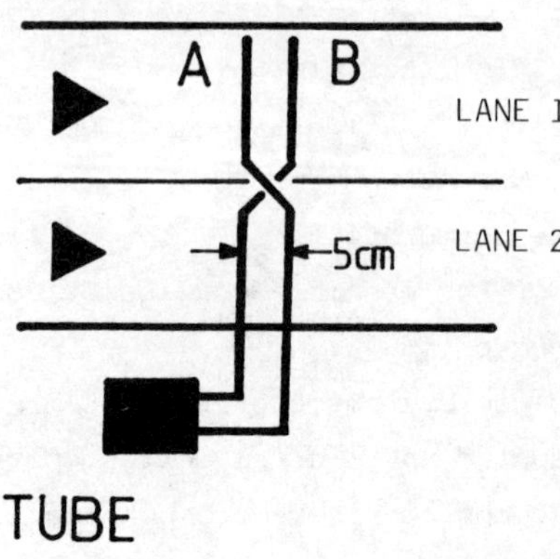

Figure 5 Lane Counting

relatively low speed of sound in air, which can easily introduce timing errors of several milliseconds. The Truvelo system, aimed at manual operation for speed limit enforcement, gets round this by using a pair of coaxial cables instead of pneumatic tubes. Lyles and Wyman (1980) and Davies (1982) present more general information on speed monitoring system accuracies.

Vehicle Type Classification

The basic methodology of vehicle type classification, as developed at the Transport and Road Research Laboratory (TRRL), utilises an array of road sensors in each traffic lane. The sensor signals are processed with the aid of a microprocessor to determine the vehicle parameters. These parameters, principally axle spacings, are compared with a table of values stored in the microprocessor. When a parameter match is obtained, the vehicle class is identified.

TRRL's Automatic Vehicle Classification System (AVCS) is based on the RCA 1802 microprocessor, and its successor, the RCA 1805. Use of the 1802 dates from 1976, when the automatic classification system ALICE was first implemented in microprocessor hardware. Dalgleish and Tuthill (1978) describe the development of the processing unit of this prototype classification system based on earlier Golden River Mk 4 traffic monitoring equipment.

Although more powerful microprocessors are now available, replacement of the RCA unit is unlikely. Low power consumption and a robust operating specification suitable for extreme environmental conditions are of more significance than processing power, and will probably remain so in the immediate future.

The AVCS may be considered as four component parts that combine to form the classification system. These are:

(a) the vehicle sensors;

(b) the vehicle detectors units;

(c) the microprocessor and its peripherals; and

(d) the microprocessor software.

Two road sensor arrays were evaluated for the AVCS, as shown in Figure 6. The preferred arrangement consists of two triboelectric axle sensors and one inductive loop per lane. Some earlier systems consisted of one axle sensor and two inductive loops, which offers some minor advantages and disadvantages by comparison. Both sensor arrays are symmetrical, so that with suitable signal processing, vehicles may be classified for both directions of travel. This is important for two-directional roads, and for other highways where traffic contraflow can occur.

For the purposes of vehicle classification, the sensor array in Figure 6A produces the more accurate results. This is primarily due to the different detection zone for each sensor array. Classification errors are observed when one vehicle is overtaking another, passing over the sensor array at short headways. The shorter the array, the less likely this is to occur.

Sensor output signals are initially processed by axle and loop detector units. The detector outputs are clean noise-free pulses, which are fed to the microprocessor. The microprocessor is programmed to inspect these output signals at one millisecond intervals. A signal time sequence diagram is shown in Figure 7,

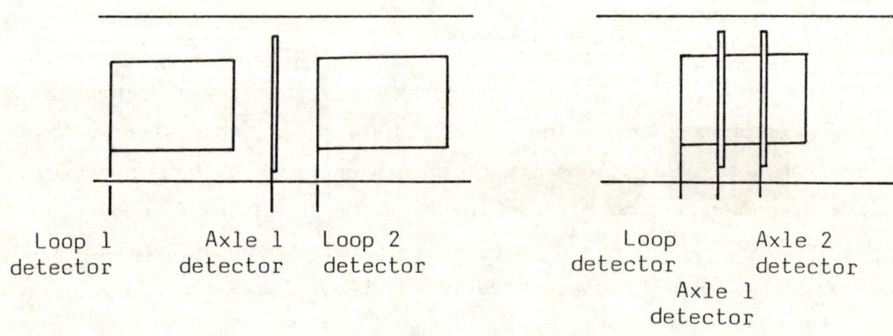

Figure 6

for a three axle vehicle crossing the preferred sensor array.

From the timing points, the microcomputer can calculate the following vehicle parameters.

<u>Speed</u> = $\dfrac{d}{t3-t2}$ kph

<u>Length</u> = Speed x (t6-t1) - L + K metres where K is a factor associated with the loop detection zone.

<u>Wheelbase 1</u> = Speed x (t4-t2) metres.

<u>Wheelbase 2</u> = Speed x (t5-t4) metres.

<u>Overhang</u> = Length - (Wheelbase 1 + Wheelbase 2) metres.

<u>Axle Count</u> This can be obtained directly from either axle detector.

<u>Inter-vehicle gap</u> = t7 - (t6 + time for 1st vehicle to travel dimension L).

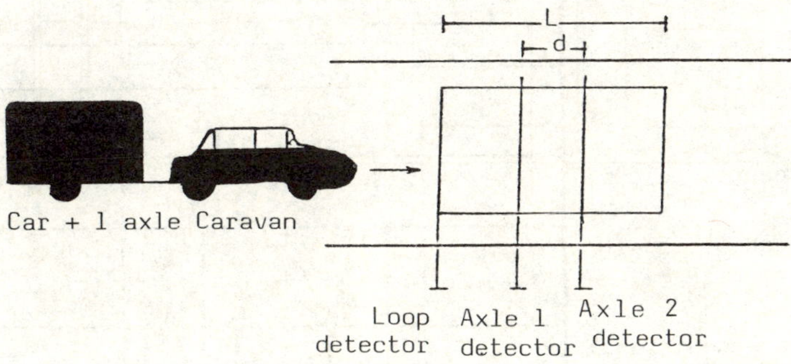

Figure 6A

Following the onset of the loop detector vehicle presence pulse at (t1) the microcomputer will store all the subsequent timing points until the vehicle has passed over the road loop at (t6). All the stored timing points are thus associated with that individual vehicle, and the microcomputer can calculate the vehicle parameters. If the first pulse transition of axle detector number 2 occurs prior to number 1, then the vehicle crossing the sensor array is travelling in the "non-normal" direction. Once this is established the microcomputer takes this into account while calculating the vehicle parameters.

Further refinement of inductive loop technology has allowed the estimation of chassis heights of passing vehicles as an additional parameter for vehicle classification. Two techniques of chassis height monitoring are feasible: in the first, an analogue voltage output from a conventional loop detector is gated through different threshold voltages, while in the second, a digital signature is processed directly by suitable software. Chassis height measurement provides a valuable additional

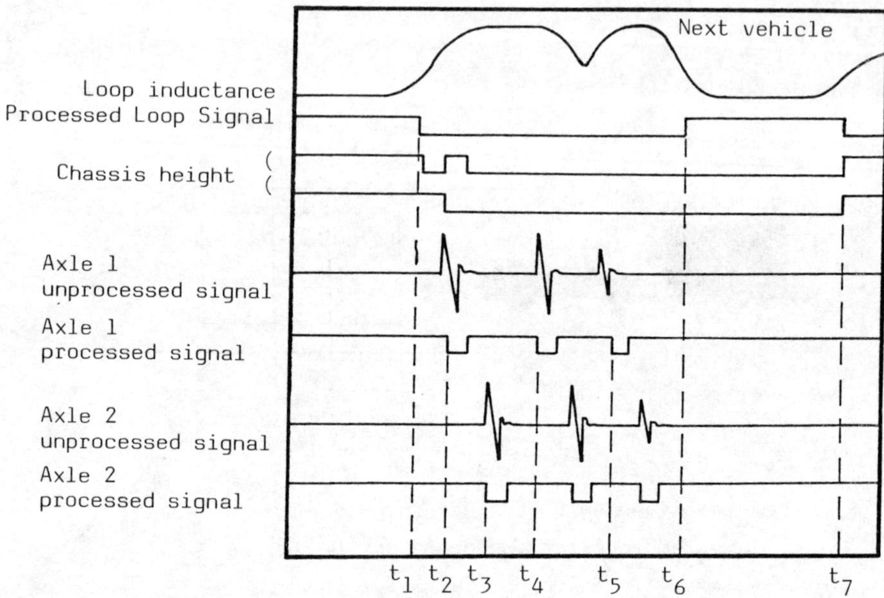

Figure 7 Sensor Time Sequence Diagram

parameter for automatic vehicle classification (Moore et al, 1982).

Research in the United Kingdom has concentrated on development of techniques and equipment which can automatically count and classify vehicles into 25 separate categories. These categories were primarily defined from the EEC requirements, regulation R1108/70. However, in addition to the EEC requirement certain additional categories were included such as bicycle, motorcycle, car plus trailer, while more recently the number of categories for the UK national core census has been reduced to 20 (Brown-Kenyon and Butler, 1984).

A major problem in the development of detailed vehicle-type classification equipment has been the lack of a consensus on the definition of appropriate vehicle categories. The US Federal Highway Administration (FHWA) produced a definitive vehicle type classification scheme in 1984, following extensive consultations with the states. State responses to the original FHWA proposals generated 80 alternative vehicle groupings, which were eventually reduced to thirteen. Individual states are still free to add further categories of their own choice, however, provided that they supply national statistics compatible with FHWA guidelines.

The vehicle classification 'Scheme F' currently recommended by the FHWA comprises:

1. Motorcycles (which are an optional category);
2. Passenger cars, including cars pulling trailers;
3. Other two-axle, four-tyre single unit vehicles;
4. Buses of all types (except minibuses);
5. Two-axle, six tyre single unit trucks;
6. Three-axle single unit trucks;
7. Four-or-more axle single unit trucks;
8. Four-or-less axle single trailer trucks;
9. Five-axle single trailer trucks;
10. Six-or-more axle single trailer trucks;
11. Five-or-less axle multi-trailer trucks;
12. Six-axle multi-trailer trucks; and
13. Seven-or-more axle multi-trailer trucks.

A fourteenth category can be added for any vehicle which does not fit into the thirteen categories listed above. Additionally, some states have defined specific sub-divisions of certain categories, resulting for example in a total of twenty-two categories recently used in the State of Iowa.

Lyles and Wyman (1982) reported that only one automatic vehicle classification system was able to satisfy classification requirements similar to those listed above; that was the system developed at the UK Transport and Road Research Laboratory (TRRL). Since then, systems have been developed by several manufacturers which meet FHWA requirements to a greater or lesser extent. An example is the Canadian IRD Classifier CL-400 which can distinguish nineteen vehicle categories and monitor between two and six lanes. This system uses two inductive loops with a magnetometer type axle sensor in each lane. These sensors are designed for permanent all-weather usage and enable classification to be made on vehicle speed, length and axle spacing. Additionally the axle sensor will discriminate between single and double tyres. Other commercial type classifiers, such as those manufactured by GK Instruments and Golden River, are generally single lane systems based on a pair of pneumatic tubes, with or without an inductive loop.

CONCLUSION

This chapter has detailed established techniques and new approaches to vehicle detection, which have resulted in the development of real-time vehicle classification systems. With the advent of microprocessor technology automatic vehicle classification is now a practical proposition, providing statistically representative traffic data with better spatial and temporal coverage than could be achieved using traditional manual techniques. These techniques and systems, which are already gaining wide acceptance, will allow better and more accurate traffic data to be collected, to the long-term benefit of all highway users.

REFERENCES

Allsop, R.E. (1984). Introduction to traffic survey methods. In Taylor, M.A.P.(ed.)Traffic Survey Methods. Department of Civil Engineering, Monash University.

Basson, J.E.B. & W.D.O. Paterson (1977). A brief discussion of the NITRR axle weight analyser. Report RP/10/77. National Institute for Transport and Road Research, Pretoria, South Africa.

Blackmore, D.H. (1966). Operation and maintenance of the Fischer and Porter punched tape counters. RRL Report SR-9 Road Research Laboratory, Crowthorne.

Brown-Keynon, A. and A. Butler (1984). Automation of the Department of Transport's core census of traffic. Proc., Int. Conf. on Road Traffic Data Collection, Institution of Electrical Engineers, London.

Byrne, B.F., R.F. Roberts, L.E. King, & R.G. Arbogast (1976). Testing of the tape switch system for determining vehicle speeds and lateral position. Transportation Research Record 615, TRB, Washington DC.

Dalgleish, M.J. and R.D. Tuthill (1978). The development of a microprocessor-based vehicle classifier. Traff. Engineering Control, 19 (3), pp 116-119.

Daniels, D.J. (1982). An experimental FMCW traffic radar. IEE International Conference on Road Traffic Signalling.

Institution of Electrical Engineers, London.

Davies, P. (1982). The accuracy of automatic speed and length classification. Department of Civil Engineering, University of Nottingham.

Davies, P. (1983). Automatic vehicle classification techniques: Lane, speed and basic type classification. Traffic Eng. and Control, vol 24(4) pp 195-201.

Davies, P., M. Bettison and F.K. Sommerville (1985). Development of a low-cost truck weighing system. Final Report to the US Federal Highway Administration. Department of Civil Engineering, University of Nottingham.

Davies, P., and D.R. Salter (1983). The reliability of classified traffic count data. 62nd Annual Meeting of the U.S. Transportation Research Board, Washington DC. Also in Transportation Research Record 905, 17-26.

Davies, P., and N.D. Ayland (1985). New developments in Weigh-in-Motion. US Society of Automative Engineers, Annual Meeting, Portland, Oregon.

Davies, P., D.R. Salter and M. Bettison (1982). Loop sensors for vehicle classification. Traff Enging Control, 23(2).

Davies, P., D.R. Salter and F.K. Sommerville (1986). Fundamental properties of piezo-electric axle load sensors. Summary of laboratory research programme results. Department of Civil Engineering, University of Nottingham.

Davies, P., and F.K. Sommerville (1985). Low-cost WIM: the way forward. 2nd National Weigh-in-Motion Workshop, Atlanta, Georgia.

Department of Transport (1981). Traffic Appraisal Manual. Department of Transport, London.

Dickinson, K.W. et al (1982). Video image processing system for gathering road traffic data. University of Manchester Institute of Science and Technology.

Dickinson, K.W. (1983). Future developments in traffic monitoring. Symposium on Highway Maintenance and Data Collection , University of Nottingham.

Duley, R.K. (1981). The loop detector and its applications. ICE Conference on Sensors in Highway and Civil Engineering. Institution of Civil Engineers, London.

Franklin, P. (1984). Automatic vehicle classification using a profile detector. IEE International Conference on Road Traffic Data Collection. Institution of Electrical Engineers, London.

Griffiths, P. (1982). The development of traffic data collection systems by the Mangood Group. Internal report. Mangood Limited., Pontypool, Gwent, Wales.

Gruen, P.J. (1985). The role of traffic monitoring in West Yorkshire. J. Inst. Highways and Transportation vol. 32(6) pp 19-23.

Head, J.R. (1982). Detecting vehicles with inductive loops and microwaves. Proc., Int. Conf. on Road Traffic Signalling. Institution of Electrical Engineers, London 1982.

Hillier, J.A., D.H. Mathews & R.C. Moore (1978). Trends in traffic data collection and analysis. ICE Conference on

Traffic Data Collection, Institution of Civil Engineers, London.
Honey, D.W. (1975). Sensing an approaching vehicle. Surveyor, 9 May 1975.
Kember-Smith, J. (1985). Temporary inductive road loops for automatic traffic recording. TRRL Application Guide AGI, Transport and Road Research Laboratory, Crowthorne.
Lenz, K.H. & R. Hotop (1974). Fully automated acquisition and processing of traffic data. Traffic Engineering and Control. Aug/Sept (1974).
Lyles, R.W. and J.H. Wyman (1980). Evaluation of speed monitoring systems. Maine Department of Transportation and University of Maine at Orono, Final report No FHWA/PL/80/006 for U.S. Department of Transportation (Federal Highways Administrations, Office of Highway Planning).
Lyles, R.W. and J.H. Wyman (1982). Evaluation of vehicle classification equipment. Maine Department of Transportation (Federal Highways Administration, Office of Highway Planning).
Mills, M.K. (1974). Magnetic gradient vehicle detector. IEEE Transactions on Vehicular Technology, vol. VT 23 (3) p 91-99.
Moore, R.C. (1981). Road Sensors for traffic data collection. Proc. Conf. on Sensors in Highway and Civil Engineering Institution of Civil Engineers, London.
Moore, R.C., P. Davies and D.R. Salter (1982). An automatic method to count and classify road vehicles. Proc Int. Conf. on Road Traffic Signalling, Institution of Electrical Engineers, London.
Moore, R.C. (1983). An automatic method to count and classify road vehicles. OECD Seminar on Road Traffic Information. In TRRL Report SR 757, Transport and Road Research Laboratory, Crowthorne.
Moore, R.C. (1984). Traffic data collection and the impact of the microcomputer. J. Inst. Highways and Transportation, vol. 31 (5), pp 2-9.
Nash, D.D. (1976). ALICE Automatic length indication and classification equipment. Traffic Engineering and Control, December (1976).
Reijmers, J.J. (1980). On line vehicle classification. IEEE Transactions of Vehicular Technology. Vol VT29(2).
Richardson, A.J. and M.A.P. Taylor (1984). The traffic survey process. In Taylor, M.A.P. (ed.) Traffic Survey Methods. Department of Civil Engineering, Monash University.
Salter, D.R. and P. Davies (1984). Development and testing of a portable capacitive WIM system. 63rd Annual Meeting, Transportation Research Board, Washington D.C.
Scarzello, J.F. et al, (1978). Development of a self-powered vehicle detector. Federal Highway Administration Report FHWA RD-78-89.
Storr, P.A., J. Wennell, M.R.C. McDowell and D.F. Cooper (1979). A microprocessor-based system for traffic data collection. Traffic Eng. and Control, 20(4), 156-158.
Taylor, M.A.P. (1984). Needs for traffic data. In Taylor, M.A.P. (ed.). Traffic Survey Methods. Department of Civil Engineering, Monash University.

Wootton H.J. & R.J. Potter (1981). Video recorders, microcomputers and new survey techniques. Traffic Engineering and Control. Vol 22 (4).

Van Helden, D. & R.C. Van der Vuurt (1974). Automated analysis of highway traffic. Traffic Engineering and Control. August/September 1974.

EDITORS' NOTES

(1) Optical detectors have recently been installed in Japan on the route between Tokyo and the site of the World Exposition. Details are given in a paper by Tsugawa, Yoshida, Odake and Maruta entitled "Traffic control system and new type sensor at Tsukuba Exposition 85" in the proceedings of the Second International Conference on Road Traffic Control, Institution of Electrical Engineers, London, April 1986. A mask is placed over the photoelectric element array behind the lens of the camera so that a number of windows are defined in the field of view. The passage of vehicles through these windows causes variations in the light intensity which can be detected electronically. In this way, the optical detector can replicate one or more on-street vehicle detectors. High levels of detection accuracy both during the day and at night are claimed. For a further discussion of vehicle detection by image processing see Chapter 4.

(2) For further details, see Chapter 4.

(3) It is interesting to note that there are distinct national preferences for different types of vehicle detector. For example, inductive loop detectors are the predominant form in the UK whereas ultasonic devices are generally preferred in Japan.

Information Technology Applications in Transport, pp. 41–64
P. Bonsall, M. Bell (Editors)

Chapter 3
AUTOMATIC VEHICLE IDENTIFICATION

Ian Catling
Transpotech Ltd., UK

INTRODUCTION

Automatic Vehicle Identification (AVI) is the term used to describe systems which enable vehicles to be uniquely identified at various locations on a network. This chapter discusses the various uses of AVI and the different technologies which have been applied with varying degrees of success over the past two decades. It concludes with a description of the Hong Kong Electronic Road Pricing (ERP) pilot stage project which was successfully completed in 1985, and which included the first large-scale on-street demonstration of AVI.

AVI is the logical next step after automatic vehicle detection and automatic vehicle classification - an AVI system not only detects a vehicle and its type, but identifies which particular vehicle is at a particular place at a particular time. The potential uses of AVI systems are numerous, and range from improvements in public transport facilities through better fleet management techniques to the economically efficient control of congestion by means of road pricing.

ELECTRONIC ROAD PRICING

It was research into road pricing in the late 60s and early 70s, following the Smeed report of 1965, which generated the first (1) serious and practical research into methods of achieving widespread

AVI, although as described below there had been earlier experiments mainly for rail-based systems. The theory of road pricing is that road users should pay for their use of road space according to how much they are contributing to congestion. In other words, at off-peak times, or on uncrowded roads, there should be no special charges, but in town centres which are congested at peak times there should be charges imposed for those road users who choose to use those roads at those times. By setting the charges correctly, the most economically efficient use of the road network can be achieved. Most trips will still be made in the same way and will benefit from the reduction in congestion. Some users however will have changed their time of travel in order to avoid the higher charge, or have changed the mode of their trips, or have modified their behaviour in some other way - the removal of these marginal lower value trips from the most congested roads produces a very significant net benefit to the community.

Road pricing therefore depends upon charging vehicles (or their owners) for being in a particular place at a particular time. A form of road pricing is the Area Licensing System which has been in force in Singapore since 1975. Vehicle users buy daily or monthly tickets which are displayed on the windscreen and are checked during peak hours at the 27 entry points to the city's Central Business District - the 'restricted zone'. The ALS is a form of supplementary licensing, which was seriously considered for introduction in London around the same time, and which has proved to be extremely successful in controlling congestion within the restricted zone during the morning rush hour. But it is not very flexible - it would be administratively very difficult for example to introduce a second zone, let alone a third or fourth, and there is no question of charging by time period other than the single peak period. In order to introduce this level of flexibility, an AVI system is needed so that a record is kept centrally of the charges incurred by each vehicle, and bills can be sent out at regular intervals to vehicle owners. A whole series of charge points - whether forming zones or not - can then be established, and the charges can vary by place and by time of day. This is the basis of the Hong Kong system described at the end of this chapter. Early suggestions for implementing road pricing using on-vehicle equipment such as meters or charge cards have been largely ruled out on security grounds.

USES OF AVI

As the technology of AVI began to evolve during the early 70s, the first practical uses of it were made in rail-based systems. This was because it is much less demanding of the equipment to identify a vehicle which is guaranteed to be moving always in exactly the same direction, in exactly the same lateral position across the track and at exactly the same height off the ground than it is to deal with multiple vehicles, weaving, overtaking and the many other complications of road-based traffic. The earliest systems were often optical ones which were unlikely ever to be suitable for road use but nevertheless laid the foundation of the monitoring and control systems in use today. The Glasgow Underground system for example uses AVI to monitor centrally the exact location of each of the trains on the network, and to control the signalling accordingly. Similar systems have been in use for rail-car identification, routing and switching systems in the USA for most of the last decade. (2)

Traffic applications

One of the most obvious potential uses of AVI in road-based systems is for toll collection. The cost of building toll plazas and the associated toll collection systems, together with the operating costs of the usually highly labour-intensive collection operations, means that potential increases in efficiency are always under scrutiny by toll authorities. This is nowhere more so than in the USA where there have been a number of investigations in recent years into the socpe for introducing AVI to provide automatic toll collection. The idea is that regular users of the bridge, tunnel or highway would have their vehicles equipped with the necessary identity unit and would not need to stop to hand over cash when driving through the toll plaza - they would be automatically identified and would be billed regularly for their use of the facility, perhaps once a month. Such systems have been tested, but not yet generally introduced, by several toll authorities - including the New York Port Authority, the New Jersey Turnpike and the Golden Gate Bridge. In Hong Kong a decision was taken in late 1985 to use the pilot stage ERP equipment for introducing automatic toll collection at the three tolled tunnels in the territory.

The concept of automatic toll collection could readily be extended to car parking. Regular users of particular car parks, or of chains of car parks, could have their vehicles fitted with AVI identifying units and car parks could be equipped with identification sites at entries and exits. Charges could then be accrued automatically during a month according to the exact time spent in each car park, and the bill sent to the owner.

An early use of AVI was for selective vehicle priority at traffic signals. Systems were developed in Europe and in the UK which would enable traffic signal controllers to give priority to selected vehicles - usually buses, or emergency vehicles. Initially such systems were restricted to a small number of identities, but the one recently installed on buses in London, for example, enables each vehicle to be uniquely identified. Although the system is currently being used only for selective priority, it is planned to extend the scope of the system later to make more use of the fleet location information. The same principle of selective vehicle priority at traffic signals applies equally to emergency vehicles.

Fleet management applications

Information for fleet managers on the location of their vehicles is another significant potential use of AVI. Command and control systems generally depend solely upon radio communication with drivers which means that locational information is very often inaccurate and that expensive radio channels are used inefficiently. Direct connection to an AVI system can mean that the fleet manager has real time information on the location of each vehicle in his fleet, and can re-schedule accordingly. Alternatively, an AVI system could produce historical records of a fleet's usage of a road network, which would be useful for cross-checking with drivers' records, route checking, longer term re-scheduling and planning.

Fleet managers could use an AVI system for security access to depots and other facilities. As well as controlling the entry facility, the system would also record the identities of all vehicles entering or leaving. This application clearly has potential uses in many situations, not only in the commercial world.

A further application for fleet managers, and potentially for private motorists, is for vehicle refuelling. Investigations have already been carried out in some cities for the automatic recording of fuel supply with vehicle identity information to provide further data for fleet managers, particularly of bus fleets. By extension however, it would be quite possible for the private motorist to drive over or past an identifying unit as he approached a fuel pump at a garage, and he could be billed automatically - a system even simpler than the ones currently being introduced using credit cards.

As well as the vehicle performance monitoring possible from the historical record of vehicles' locations from an AVI system, public transport operators could use similar information for monitoring service performance. Much effort is currently expended by bus companies in collecting information on the level of service being provided, and much of it is inevitably somewhat inaccurate. AVI could give precise information on headways, bunching, delays and general level of service.

Another application to bus fleets is the provision of information to passengers at bus stops. In the UK a 'talking bus stop' system has been installed to provide information on the length of time until the next bus. It relies upon recognising buses upstream of the bus stop, but does not necessarily identify the route number. With a true AVI unit on the bus, and a central system controlling the flow of information, the length of time expected for buses on individual routes could be made available at bus queues. Similar systems are already in operation on a number of underground train systems.

(3)

Traffic data collection

Traffic data collection is an area which would certainly benefit substantially from many AVI systems. For instance, in a full road pricing scheme in which all vehicles permitted into an area were fitted with indentifying units, not only would accurate classified counts be available by whatever time period was required, but 'origin-destination' information normally only obtained from expensive roadside interviews could be produced as a matter of course. It has been suggested that the costs of traffic data collection are so high in some cities as to justify the introduction of an AVI system, at least on a sample basis, for data collection alone.

A major project in the USA is investigating the benefits of an AVI system for heavy goods vehicles combined with weigh-in-motion technology. In addition to the information which could be provided to the fleet owners and managers themselves on the location of their vehicles, the proposed system would provide the authorities with valuable information for enforcement and monitoring of weight, height and length regulations, and for the collection of weight/distance taxes.

An important secondary use of many AVI systems is improved vehicle security and protection against theft. In a road pricing application for example, in which all vehicles are normally fitted with identification units, any vehicle which had been reported as stolen would be detected by the system as soon as it was driven over an identification site. Even in a system without universal coverage, high value and security vehicles could be fitted with identification devices to ensure that they were identified at each site.

AVI TECHNOLOGY

Recent advances in technology based on the advent and evolution of the microprocessor have seen increasing reliability and decreasing costs, bringing the genuine possibility of introducing AVI for any of the purposes described above. The Hong Kong road pricing system in particular, described later in this section, has shown for the first time in a major trial that those applications dependent upon secure and accurate identification under all types of road condition can be implemented. The next few pages of this section discuss the various technologies which have been tried, with varying degrees of success, in arriving at this position.

Almost all AVI systems share the same fundamental design which consists of three basic components - a vehicle-mounted identification unit which contains, in some form or other, a code unique to the vehicle (or at least to the class of vehicle); a roadside or trackside reader unit which is capable of reading the identity in the unit of any vehicle passing over or by the reader; and a central processing system which records or analyses the vehicle location information sent back by the reader units. (4)

Optical and infrared systems

In the mid 1960s some of the earliest AVI systems were developed using optical recognition techniques, mainly for rail based applications. Bar codes or other labels, usually manufactured from highly reflective materials, were attached to the side, or sometimes the windscreens, of vehicles. A scanner unit, often using low power laser technology, would read the information as the vehicle passed and decode the vehicle's identity.

Such systems suffered from the requirements of good visibility and of line-of-sight contact between vehicle and scanner. Rain, snow, fog, ice and dirt all seriously degraded system performance, and while the line-of-sight requirement is often acceptable for rail systems it is a significant drawback for many road-based applications - any multi-lane traffic poses problems of occlusion. In addition the systems were sensitive to label or scanner misalignment, focussing problems and depth of field limitations, and usually would only operate at low vehicle speeds. Although recent advances have meant that the technology is well suited to more controlled environments such as store checkouts, optical systems are now generally ruled out for AVI.

In the early 1970s an extension of the optical systems was developed using infra-red. In some systems the same principle of reflection was used - sometimes by embedding an infra-red reflective material in a vehicle's windscreen - and in others a transmitter was placed on or in the vehicle. Such systems however shared most of the disadvantages of optical systems and are not now seriously advocated for AVI.

Radio and satellite systems

Perhaps the most common type of vehicle location system depends upon radio communication between driver and control centre. With the advent of dead reckoning navigational systems connected to compass and odometer together with advances in radio communication, particularly cellular radio, such systems are feasible for use by small fleets of vehicles. Satellite location fixing systems are likely to become available within the next few years which are (5) economically viable and extremely accurate. There has been some development work done to make the transmission of the vehicle's location automatic, thus achieving similar results to a more conventional AVI system. However the use of multiple radio frequencies and the lack of security features make such systems unsuitable for many of the applications of true AVI described above.

Inductive loop systems

In the early 1970s research began to be carried out, at the Transport and Road Research Laboratory amongst other places, on the use of inductive, or electromagnetic, coupling between a vehicle mounted transmitter and roadside or trackside equipment usually by means of loops buried in the road or track surface. The principle is an extension of the use of inductive loops for vehicle detection, which was already well-established. The signal (6) transmitted by the vehicle unit modulates the carrier signal in a receiver loop in order to indicate a binary digit (bit). By receiving a series of bits according to the identity stored within the vehicle unit, the roadside equipment connected to the loop can decode the identity of the vehicle.

Early inductive loop systems used a very small number of identities, some as few as four. Whilst this clearly did not give unique vehicle identification, such systems were useful for the sort of selective vehicle priority application at traffic signals described above. Later systems however were able to transmit reliably several bytes of data which is enough for any AVI application - the Hong Kong system for example allows for over 16 million unique identities as well as for two bytes of variable data (bus route, occupancy, driver ID, load information, for example).

The nature of the inductive system means that the obvious place for mounting the vehicle unit is underneath the vehicle. During the development of the early systems it soon became clear that the power supply for the vehicle-mounted transmitting unit was a potential problem, particularly in some rail applications where trucks might not be carrying power themselves, or for heavy goods vehicle trailers whose power would only be active when the vehicle's lights were switched on. Two distinct types of inductive loop system have developed - active systems, in which the vehicle unit is powered either by a connection to the vehicle's own power supply or by an internal battery, and passive systems, in which the unit is energised by power transmitted from another inductive loop in the surface beneath the vehicle.

Active inductive loop systems based on connection to the vehicle's power supply have been successfully applied to bus identification in Europe, the USA and Australia. Recent improvements in battery technology have meant that it is possible to use internal batteries to power vehicle units for transmission when they are triggered by inductive loops. For many applications the reliability of such systems is adequate - the system fitted to London Regional Transport buses for example uses internally powered units - because relatively high failure rates after a few years can be tolerated within a single fleet or a small number of fleets.

However, for an application such as road pricing in which the costs of re-fitting large numbers of vehicles with replacement batteries or whole units is significant, and in which it is therefore desirable for AVI unit life to be longer than vehicle life, battery technology does not yet provide a high enough degree of reliability. A more appropriate approach is the passive inductive loop system, in which electromagnetic induction, similar to that between the primary and secondary coils of a transformer, is used to provide power to the vehicle unit. The unit itself therefore contains no power source nor direct connection to one, and can hence be less vulnerable to outside interference or damage, whether accidental or deliberate.

The drawback of the passive system is the complexity of the array of loops which is needed at any site in order both to provide power and to receive transmitted identities back from the vehicle units. However, it is equalled by the complexity caused by providing a number of receiver loops to deal with multi-lane traffic. The Hong Kong project has demonstrated that both factors can be overcome and that the passive inductive loop system can be both fully operational in a harsh environment and can provide the robustness and security necessary for a viable system which depends upon some form of pricing, and hence contains inducement to fraud.

Microwave systems

The second main technology which has emerged from the multiplicity of research carried out during the 1970s is that of microwave transmission. Microwave systems (and some similar systems based upon the lower radio frequencies) depend upon a vehicle-based transmitter which when triggered by a reader unit will send its identity code to the reader using one of a variety of modulation techniques. As with inductive loops, both active and passive systems have been developed and demonstration systems have been implemented in a number of countries. Most applications have been rail-based, but a number of road applications, mainly for buses, have been implemented.

Microwave offers a potential advantage over inductive loop systems because the transmission frequency is much higher and hence more data can be transmitted to the reader unit. This could open up the scope for more sophisticated systems, such as driver or route information systems, which may be dependent upon the transfer of more data than an inductive loop system is capable of handling. (7)

There has been significantly less research and development carried out on microwave systems however than on inductive loop systems. Existing microwave systems tend to be over-sensitive to reader and vehicle unit alignment, to focussing problems and to extremes of temperature and vibration. Currently the older loop-based technology is much more reliable and secure.

The main disadvantage of microwave systems however, at least those that are currently available, is that they share the line-of-sight requirement of many of the other technologies. Vehicle units are normally mounted on the side of the vehicle, and reader units at the roadside are liable to be obscured in multi-lane operation.

To overcome this major drawback, it is proposed that arrays of reader antennae can be embedded in the pavement surface in the same way that loops are. This is possible at the lower end of the microwave frequency spectrum, but is likely to produce problems of interference between lanes which could take a significant amount of further research and development to overcome.

Image processing systems

All the AVI technologies discussed so far share the common requirement of a vehicle-mounted unit containing the vehicle's identity in an encoded form. The cost of such units is likely to form a significant part of the cost of any installed AVI system, and in many cases will form the largest single component of the total system cost. One technology which does not have the requirement for vehicle units - instead relying on the unique identity of the registration mark already displayed on every vehicle's number plate - is that of image processing. A number of projects have been carried out to establish techniques for decoding vehicles' registration numbers directly from digitised video pictures taken automatically at the roadside. A recently completed project run by the UK Home Office for example demonstrated that success rates could be as high as 70 per cent, and that this fast-moving technology could contribute significantly to stolen vehicle detection. (8) However, for many AVI applications - particularly those involving any form of charging such as toll collection or road pricing - the success rate is not likely to improve sufficiently, at least in the immediate future, to justify the use of image processing.

Other systems

Other early methods suggested and experimented with for AVI depended upon the use of permanent magnets, ultrasonic acoustics or even radioactive particles. All of these however had significant drawbacks such as alignment, speed, low capacity or reliability, and none have since been seriously pursued.

AVI IN HONG KONG

At the time of writing, no AVI technology has been demonstrated in a large-scale on-street demonstration in a wide range of road conditions other than the inductive loop system used in the Hong Kong pilot stage ERP project. The remainder of this section gives an overview of that project, and of the AVI system used most successfully.

The project ran for just over two years from 1983 to 1985, and its objectives were to demonstrate the viability of a full ERP scheme for the whole territory. They therefore embraced not only the technological elements but also a major transport study to assess the effectiveness of ERP in controlling congestion, and a complete investigation of the administrative, accounting and legislative aspects of introducing a full ERP system.

One of the primary objectives of the pilot stage project however was to demonstrate the technical viability of an AVI system suitable for the rigorous requirements of a road pricing system, with the fairly obvious differences in levels of security necessary when money is involved compared with, say, a bus priority or management system. Previously, no practical system had been demonstrated capable of reliably identifying vehicles on a road where there are the problems of weaving vehicles, multiple simultaneous transactions, many different types of vehicle and a generally much less benign environment than a railway.

However, the advances in micro-electronics during the early 1980s had made it clear during the feasibility study in Hong Kong that the reliability required might now be achievable at a reasonable cost. The AVI system chosen was a development of the inductive loop system manufactured by Plessey Controls Ltd and already in use in rail-based applications. The system had not yet been applied on the road however, and the particular problems of multi-lane operation and of inductively powering the vehicle units were the major ones faced during the development of the system.

The pilot system was as far as possible a complete subset of a potential full system, and as well as demonstrating the successful operation of the technology the project concluded with an outline design of such a full system. In addition to the data capture and validation system, a demonstration accounting system was supplied to show how bills might be produced in the full system, and a specification was drawn up for a full accounting and administration system which would be necessary for the successful running of the full system.

Figure 1 shows how the ERP system works. It consists of the three basic components described earlier - the vehicle unit containing the unique vehicle identity; the loops and roadside equipment which provide power to the vehicle units and receive and decode the identity code transmitted by the vehicle unit; and a control centre containing processors for validating and storing charge information and for producing bills for road usage for vehicle owners.

The vehicle unit is a small, inexpensive and extremely tough solid state device called an 'Electronic Number Plate' (ENP). It is attached in about five minutes to the underside of each vehicle, as shown in Figure 2, using inert gas welding. The ENP is a sealed unit, about the size of a video cassette tape, weighing approximately 1200 grammes. It contains no power source and does not need connection to any external devices, and will normally operate only when crossing a toll site. Power is supplied to the ENP from a special power loop in the road which causes the ENP to transmit its coded signal containing its unique identity. The data transmitted include security features which are designed to ensure system integrity. Each ENP is coded with a unique serial number during manufacture, which is independent of any vehicle registration number. The ENP is designed to operate without maintenance for at least 10 years.

The ENP consists of two custom designed and manufactured integrated circuits (ICs), two standard ICs, a small number of

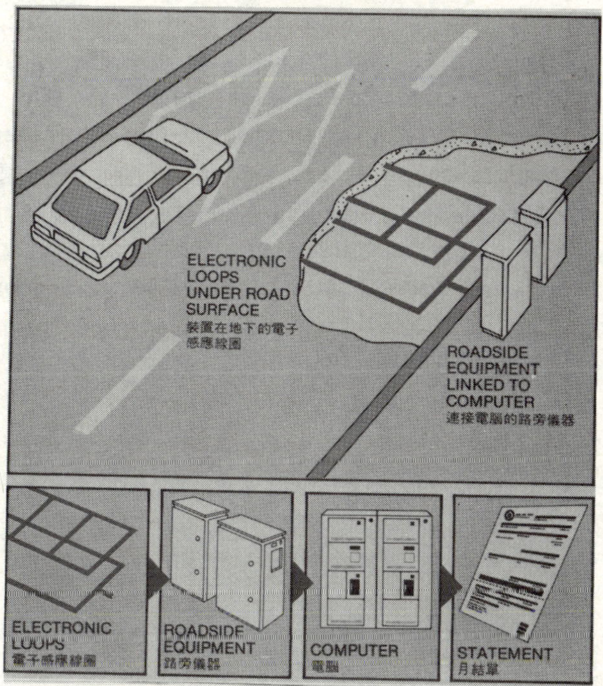

Figure 1 - System overview

Figure 2 - ENP Fitting

discrete components and 3 ferrite aerials for the reception and transmission of signals. The components of the ENP are contained in a tough, moulded plastic case which, after the ENP has been tested, is completely filled with an epoxy resin which encapsulates all of the components and provides protection against shocks and vibration - it also makes tampering with the circuits extremely difficult, and any attempted tampering very obvious.

A series of 'charge zones', which for the Hong Kong urban area would comprise up to 200 sites, are defined and motorists are charged for each zone boundary crossing during busy times. At each road which crosses a zone boundary there is a toll site, an example of which for a three-lane carriageway is shown in Figure 3. A common power loop is laid across all lanes at a toll site, capable of energising any ENP crossing the site. A number of separate receiver loops are laid across the site, overlapping the power loop and each other. A normal configuration would have two receiver loops per physical lane, to ensure that no vehicles were missed due to any lack of lane discipline.

Roadside cabinets similar to those used for traffic signal control contain a number of microcomputers handling data from the receiver loops. The outstation contains a loop driver, to provide power to the loops and hence to the ENPs, an interrogator to decode the signals from the receiver loops, and an outstation transmission unit (OTU) to check the data from the interrogator and communicate with the control centre. The interrogator comprises a set of receiver cards and a vehicle detector card. Its purpose is to pass valid, coded data from ENPs, which is syntactically correct, to the outstation transmission unit (OTU) for analysis. Each receiver loop is connected to a receiver card which checks the messages transmitted by any ENP in the vicinity of the loop and passes it to the OTU. The detector card is required only to provide some of the enforcement facilites of the system - it is connected to all of the receiver loops and monitors them for the resence of a vehicle, passing this information to the OTU which then correlates the ENP and detector information to detect any unfitted vehicles or faulty

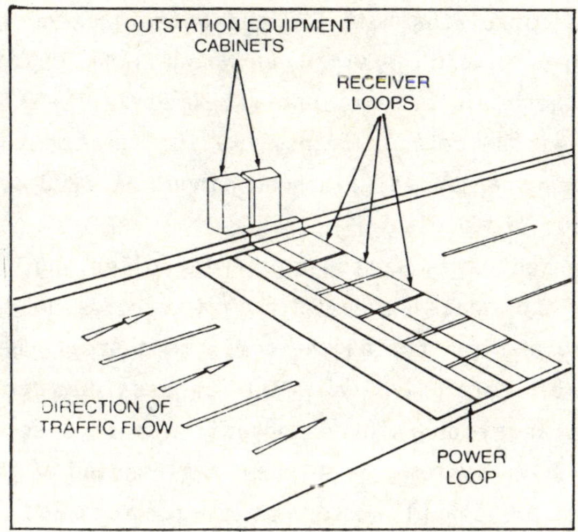

Figure 3 - Toll site layout

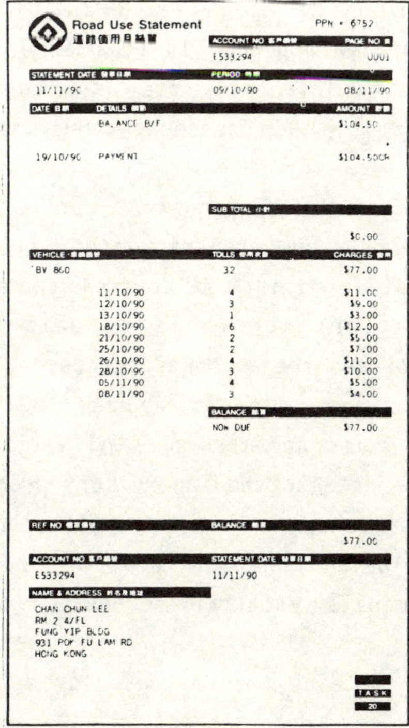

Road Use Statement

PPN • 6752

ACCOUNT NO	PAGE NO
E533294	0001

STATEMENT DATE	PERIOD	
11/11/90	09/10/90	08/11/90

DATE	DETAILS	AMOUNT
	BALANCE B/F	$104.50
19/10/90	PAYMENT	$104.50CR

SUB TOTAL

$0.00

VEHICLE		TOLLS	CHARGES
BV 860		32	$77.00
	11/10/90	4	$11.00
	12/10/90	3	$9.00
	13/10/90	1	$3.00
	18/10/90	6	$12.00
	21/10/90	2	$5.00
	25/10/90	2	$7.00
	26/10/90	4	$11.00
	28/10/90	3	$10.00
	05/11/90	4	$5.00
	08/11/90	3	$4.00

BALANCE

NOW DUE $77.00

REF NO	BALANCE
	$77.00

ACCOUNT NO	STATEMENT DATE
E533294	11/11/90

NAME & ADDRESS

CHAN CHUN LEE
RM 2 4/FL
FUNG YIP BLDG
931 POK FU LAM RD
HONG KONG

TASK
20

Figure 4 - Example road use statement

ENPs. The OTU correlates both ENP and detector signals, detecting unequipped vehicles and determining which lane a fitted vehicle occupies. It accumulates data until polled by the control centre equipment. Then the data is transmitted to the control centre via modems and four-wire leased telephone lines, at 1200 baud using the HDLC synchronous protocol.

The control centre is responsible for collecting, checking and storing all of the data received from the roadside equipment and for controlling and monitoring the complete system. Because of the large amount of data involved, the process has to be largely automated but manual operator intervention must also be possible. Traditionally such systems have been implemented with a powerful minicomputer or small mainframe equipped with suitable communications and operator facilities. Because system reliability is usually important a duplicate computer is almost invariably required. The advent of small, powerful microcomputers has permitted an alternative approach to be used in the road pricing system which has many advantages over the conventional design. This is the use of multiple processors, connected together using a local area network.

The primary function of the control centre processors is to validate vehicle data before preparation of bills by the accounts processor. Processing within the control centre is distributed between communications controllers, data validators, the supervisory processor and the accounts processor.

Each communications controller is responsible for data collection from up to 48 toll sites and then passing valid ENP data to data validators and other data to the supervisory processor. Each toll site is polled at least every 6 seconds, with those connected to a CCTV outstation or being monitored from a local terminal polled more frequently (nominally each half-second, depending upon loading on the system).

A data validator contains the database for a discrete subset of the whole vehicle fleet. Upon receipt of ENP data from OTUs (via a communications controller), it checks the identity of each vehicle therein before sending the accounting information to the accounts processor. Whenever a DV fails to validate a vehicle, it passes the data to the supervisory processor for further (anomaly) processing and, if the data originated at a CCTV or monitored site, a message is sent to that site to recover a photograph of the vehicle or display the information on the monitoring terminal.

The functions of the supervisory processor are to manage the system, supervising the databases and state of all other processors; to process "anomalies", or apparently illogical transactions which have not been validated by a data validator; and to enable operators to observe the system and monitor vehicles and toll sites.

Validated data is accumulated during the month, and the accounts processor produces for each owner a statement of his road use charges which is sent to him at the end of the period. The bill is similar to a credit card statement, and in Hong Kong will be payable by a number of methods familiar to vehicle owners. A typical statement is shown in Figure 4.

A CCTV system, supplied in the pilot scheme by GEC Avionics, with cameras installed at selected sites ensures that any vehicle without an ENP - or one whose owner is trying to cheat the system - is photographed. The pictures are transmitted to the Central Office where appropriate enforcement action is automatically initiated.

Initially one of the major criticisms of the proposed ERP system was that the security aspects would not be sufficiently robust to deter widespread evasion of road pricing charges. During the course of the project however, it was successfully demonstrated that the enforcement aspects of the proposed system were more than sufficient to make undetected evasion all but impossible and the certainty of being caught so high as to make the expected level of attempted evasion well within administratively manageable bounds.

The system incorporates strict controls on access to the vehicle data collected and ensures that no permanent record of a vehicle's movements is kept. Although records of individual transactions will be kept until after the appropriate charges have been paid, there is of course no record of individual trip patterns or of who was driving a vehicle - the CCTV photographs are specifically designed for number plate recognition and do not include the driver of a vehicle.

Figure 5 is a map of the Central district of Hong Kong Island, showing the 18 on-street sites which were installed for the pilot stage (two off-street sites were used for commissioning and testing). The sites defined a 'watertight' zone through which any vehicle entering Central had to pass, and the test sites within the zone ensured that most movements through the system generated three or more transactions.

Some 2600 vehicles were fitted with ENPs for the pilot stage. About 1200 of these were Government vehicles, 700 or so were buses and the remainder belonged to volunteers - companies and individuals - who regularly use the area. A wide range of vehicle types was included and there were no difficulties in fitting the ENP quickly and simply with an average fitting time of around 5 minutes.

The pilot stage data capture equipment was a complete subset of a potential full system. Because of the modular nature of the control centre computers, the design and implementation of each part of the pilot stage system is equally appropriate to a full system.

A set of administrative procedures was developed to deal with all aspects of running the ERP system, including:

- the provision, testing, replacement and removal of ENPs,
- maintaining accurate databases of vehicle ownership,
- statement production and distribution,
- accounts maintenance and reporting,
- audit requirements,
- payment methods,
- enquiries,
- staff functions and requirements, and
- all aspects of enforcement including debt collection and

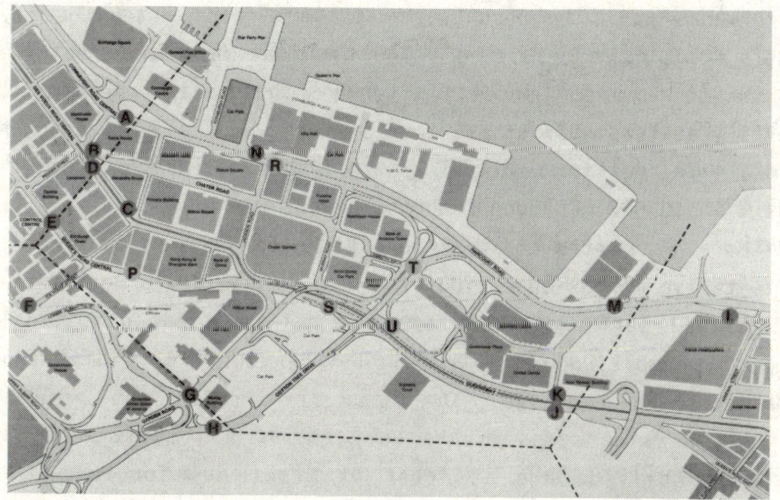

Figure 5 - pilot sites in Hong Kong

procedures for dealing with suspect vehicles detected by the data capture system, the CCTV enforcement system or the police enforcement team.

The overall system performance measurements were perhaps the most important to be taken during the pilot stage, as it was largely these results that determined whether or not a full system was feasible. If the requirements of the performance specification had not been achieved, it might have been argued that improved performance would be possible from a modified system design, but doubts would have persisted. The two most important requirements of the performance specification were that the system should identify at least 99% of all four or more wheeled vehicles crossing a toll site in a form suitable for billing and ensure that not more than 1 in 10,000,000 such identifications was erroneous.

After eight months of operation of the complete pilot system (and up to twelve months operation of some parts), it was confirmed that the overall performance is more than adequate for a full system, the system working reliably reliably under a far wider range of conditions than expected. The chances of a driver not being billed for a toll site crossing are very small, and the chances of an incorrect billing are at least as remote as for other public utilities.

The results of the pilot stage prove that the technology of electronic road pricing can operate reliably in the harsh Hong Kong environment. The design produced at the start of the pilot stage has proved very robust and has no serious shortcomings, a fact emphasised by the early installation and commissioning of the system, which permitted a much longer evaluation period than originally anticipated. All components of the design (including the system software) worked well, meeting or exceeding their specified performance, and most could be used without change in a final system, although the pilot stage evaluation has identified some useful minor enhancements.

With the parallel conclusion from the transport studies work that direct road pricing offers an equitable and economically efficient solution to the very real problem of congestion in the urban areas of Hong Kong, the pilot stage ERP project concluded that on transport, administrative and technological grounds electronic road pricing could be a viable and desirable component of the territory's continually evolving transport policy. However, during the intense public debate which accompanied the project, in which there was a widespread appreciation of the proposals, a significant body of public opinion was revealed which was strongly opposed to any form of traffic restraint and to ERP in particular. Similarly, the urgency of any decision to proceed to a full system was questioned. The Government therefore announced at the end of 1985 that for the time being any decision to proceed to the full ERP system will be shelved, to be reviewed when the level of traffic congestion, or the number of private cars on the road, shows a significant increase. In the meantime however, consultants are investigating with Government the best method of using the pilot stage ERP equipment to install automatic toll collection facilities at each of the territory's three tolled tunnels.

CONCLUSION

AVI is an exciting and relatively young application of modern technology which offers a wide range of benefits in many areas of transport. At the start of this century there was much public debate when Governments suggested and implemented the registration of vehicles with unique numbers, and required that these numbers should be displayed on the vehicles themselves. By the end of the century we are likely to see the widespread extension of vehicle identification into modern technology, and the commonplace acceptance that vehicles should carry their identities for electronic, as well as visual, recognition.

EDITORS' NOTES

(1) The Smeed Committee Report (Ministry of Transport, 1964), entitled "Road Pricing: The Economic and Technical Possibilities", London, HMSO.

(2) For an example of such a system in use on British Rail see Chapter 7.

(3) For a fuller discussion of the talking bus stop and similar systems, see Chapter 13.

(4) Systems embodying an alternative configuration wherein the reader unit is located on the vehicle and the identification unit is at the roadside are currently under development. As a method of road pricing, this configuration has advantages relating to revenue collection (through, for example, prepayment schemes) and privacy (through the absence of centralised data acquisition).

(5) Chapter 14 discusses this technology further in the context of in-vehicle navigation systems and Chapter 8 describes its role in the navigation of aircraft.

(6) For a comprehensive review of this technology, see Chapter 2.

(7) For a further discussion of the appropriate balance between the storage of information at the roadside (with transmission to the vehicle) and the provision of the information in the vehicle itself, see Chapter 14.

(8) Some details of this system are given in Chapter 4.

Information Technology Applications in Transport, pp. 65–86
P. Bonsall, M. Bell (Editors)

Chapter 4

IMAGE PROCESSING FOR TRAFFIC MONITORING

Robert Ashworth[1], Keith Dickenson[2] and Roger Waterfall[3]
[1]*University of Sheffield, UK;*[2] *Napier College, UK;*
[3]*UMIST, UK*

INTRODUCTION

With the availability of more sophisticated methods of traffic data collection, increasing demands are being placed on traffic engineers to provide better quality data for a variety of purposes connected with the planning and design of highways. Traffic flow parameters are also needed for various other applications, such as in surveillance and control systems and in connection with general road safety matters.

Vehicle detection and classification systems which make use of inductive loops embedded in the carriageway are suitable for situations where traffic data need to be collected on a continuous basis or sampled at frequent intervals of time. However, they are less suitable for data collection which requires only a relatively short observation period because of the disruption to traffic resulting from the installation of carriageway-based detectors. This applies even where stick-down loops or other forms of surface sensors are employed. Furthermore, for situations which require the monitoring of traffic within a defined area rather than at a single point on the road network, for example turning movements at junctions, several sets of carriageway-based detectors with associated logic to analyse the output would be necessary to trace individual vehicle paths through the junction.

To overcome these sorts of problems, manual rather than automatic methods of traffic data collection are often used. However, this may require a large team of enumerators, with consequent resource implications, and for this reason photographic methods of data collection are now frequently employed, using an elevated camera (2) position to obtain coverage over a wide area. In recent years, videotape recording of traffic scenes has gradually replaced the use of conventional photography, particularly since the introduction of low-cost portable VTR systems capable of storing several hours traffic data on a single cassette. In addition, the availability of slow-motion and freeze-frame playback features allow accurate abstraction of event timings from a digital clock superimposed on the video recording.

The main disadvantage of video recordings, in common with other photographic techniques, is simply that manual abstraction of traffic data is a tedious and time-consuming process. Typically it takes an operator five to ten hours to analyse a one-hour recording, depending on the complexity of the scene and the amount of information to be abstracted. To some extent, this analysis procedure can be semi-automated by linking the video recorder with a microcomputer and keying in traffic events as they occur during repeated playbacks of the recorded sequence (Wootton and Potter, 1981 and Mortelmans and Venstermans, 1984). However, the degree of manual involvement is still quite high and this suggests that significant benefits might be obtained by fully automating the data abstraction from video recordings as part of a video image processing system.

IMAGE PROCESSING SYSTEMS

A typical configuration for an image processing system is shown in Figure 1. The analogue signal provided by the camera is digitised into one of a number of grey levels (usually 16, 64 or 256 in total) for each picture element or pixel and the resulting data are stored on disk for subsequent analysis off-line. With high resolution systems, each picture may comprise 512 x 512 or 1024 x 1024 pixels and with an operating speed of 25 frames/s, data are output at rates approaching 30 Mbytes/s. This precludes real-time

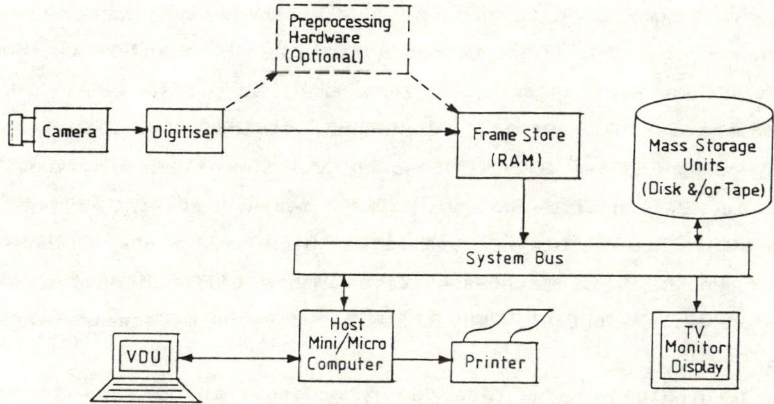

Figure 1. Block diagram of typical image processing system hardware.

analysis of the data, but is suitable for medical imagery and various types of remote sensing surveillance work on single images. At the opposite end of the spectrum, low resolution systems comprising 50 x 50 or 100 x 100 pixels with binary (i.e. black or white) images and an operating speed of perhaps 4 frames/s reduce the data output to a rate which can be handled in real-time and are therefore suitable for manufacturing processes employing robot vision. Between these two extremities lie a range of possible systems in which a trade-off exists between image resolution and frame rate in order to keep the data output to a manageable level.

Vidicon and solid-state cameras

Early image processing systems have used standard vidicon cameras which provide an analogue output (typically 625 lines, 25 frames/s with interlaced fields). This signal is then passed to an analogue-to-digital converter (ADC) to produce a digitised picture having the required number of grey levels and then, via a buffer store, to the storage medium. Vidicon cameras, however, suffer from a number of disadvantages; for example, the frame rate is non-adjustable, the signal-to-noise ratio is relatively low, the built-in automatic gain control (AGC) causes the background to darken when a light object passes across the scene and a stationary bright object such as a street lamp might be permanently burnt on the camera target. On the other hand, vidicon cameras are relatively cheap and are commonly used in CCTV surveillance systems.

As an alternative to vidicon cameras, several types of solid-state cameras are now available which have a target consisting of an array of light-sensitive points diffused onto silicon semi-conductor material. The three basic types of sensors available are:

(1) charge-coupled devices (CCDs);

(2) charge-injection devices (CIDs); and

(3) photodiode arrays.

Although they have varying properties, all can be used as imaging devices in small and robust cameras. They operate from low-voltage supplies and produce signals easily interfaced to computer systems. As each device is made up of an array of individual light-sensitive detectors, the image can be immediately quantised into pixels. The signal-to-noise ratio for each pixel is significantly better than for vidicon cameras, and since the manufacture of integrated circuits requires that the location of the detectors in the matrix is exact, high spatial accuracy is achieved.

A particular advantage of solid-state cameras for use in traffic data collection is that their frame rate is flexible. Vidicon cameras, because of their design and particularly the magnetic beam-scanning components, are tied to the mains frequency giving 25 or 30 frames/s. With solid-state cameras, varying the frame rate affects sensitivity. Thus at slow scanning rates, the integration time is long and the camera will respond to very low light-levels. However, if for computational reasons a slow frame rate is required but the light level is high, then some sensor-blooming may occur.

Sensor-blooming is caused when so much light falls on the sensor that spillage of charge to adjacent pixels takes place. This results in part of the scene becoming a large white blob or column making interpretation of the image almost impossible. Unfortunately, this kind of fault is commonly caused when observing traffic, typically by the sun reflecting off vehicle windscreens or brightwork. Headlights and reflections of lights, either from vehicles or even off wet roads, also cause trouble. Cameras are now being developed with improved anti-blooming characteristics and can incorporate electronic shutter control. Image sensors are also becoming available with on-chip processing which will enhance their flexibility for picture analysis.

Software systems and the interpretation of images

A number of ways exist for the interpretation of a sequence of digitised images and the detection of a moving object in the scene. These methods include:

(1) background frame differencing;

(2) inter-frame differencing; and

(3) segmentation and classification.

Essentially, background frame differencing is a method in which a grey-value reference image, which does not contain any vehicles, is stored and subsequently subtracted from each incoming frame or image. All non-moving features of the image therefore disappear, leaving moving grey-value objects which can be represented as binary images after applying a suitable threshold (Figure 2). However, since ambient light levels can change by an order of magnitude within seconds, it is necessary with this method to up-date periodically the stored background frame for satisfactory detection of moving vehicles over an extended time period.

Inter-frame differencing overcomes the problem of changes in ambient light levels in most cases. In this case, a background frame is again subtracted from the incoming image but the incoming frame thereafter becomes the background for the subsequent frame. The background is therefore updated with each new frame and each moving object in the scene is easily distinguished. However, systems of this sort can suffer from problems associated with matching regions of detection from frame to frame, stationary vehicles disappear and random noise is accentuated.

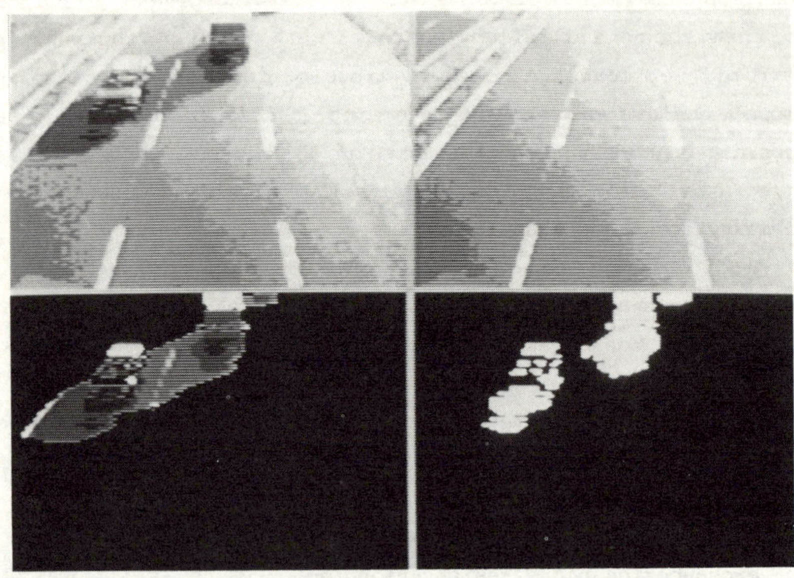

Figure 2. Processing steps in background frame differencing.
Top left : current or input image
Top right : background or reference image
Bottom left : Difference picture showing moving objects
Bottom right : binary image after thresholding and removal of vehicle shadows

Both background frame and inter-frame differencing techniques can be applied to the whole image or limited to small areas (windows) within the image. If processing is confined to interesting parts of the image, the image frame rate can be increased, giving a better measure of the time at which an event occurs.

An alternative to isolating an object by frame differencing is to interpret the raw grey-value image. This is carried out in general-purpose image processing systems by first segmenting the image into regions that are in some sense homogeneous. Classification of these regions, based on spatial features and pixel intensity, would follow and result in a name being assigned to regions exhibiting similar characteristics. Segmentation of a traffic scene into regions that might be classified as vehicles would, however, be difficult since vehicles will exhibit different shapes and variable reflectance and colour. The scene may be cluttered with irrelevant stationary objects which would also be classified and thus confuse the situation. The ability to classify a generic VEHICLE will also depend on the availability of a practical definition that distinguishes all objects that are vehicles from all those that are not.

Speed measurement and tracking vehicles

Once a vehicle has been detected in a sequence of images, several alternative methods of speed measurement and tracking can be tried; for example:

(1) measurement of the time between detection at two fixed points;

(2) estimation of the distance travelled by the vehicle between one frame and the next;

(3) cross-correlation search using grey-value signatures of individual vehicles; and

(4) feature matching.

Past and present research in the traffic field relies almost exclusively on the first two methods, since in both cases the analysis is kept simple and the processing time is minimised. The accuracy of the first method depends upon the system frame rate, but since it is only necessary to time events at predetermined parts of the image a windowing system can be adopted and frame rates can be high. In the second system, spatial resolution is most important.

In the cross-correlation search technique the vehicle has firstly to be detected by some means, after which its description, in the form of a grey-value sub-matrix of the image, is stored together with its location. When the next image is captured the 'signature' sub-matrix is correlated with the image matrix and the best-match new location is determined and recorded. The system can from past experience accommodate some change in the vehicle grey-value signature as it tracks the vehicle through a sequence of images. The major problem with this type of system is that it is computationally very expensive.

Current software developments concerned with the analysis of time-varying images suggest that by isolating prominent features on a vehicle,e.g. corners of the bodywork, more reliable tracking of vehicles could be achieved. Before this can be done, some form of edge detection must be carried out. Those edges which represent prominent features can then be matched from frame to frame and feature displacement can be determined. However, this technique is the most computationally demanding and at present is at the fundamental research stage (Nagel, 1983).

SUITABILITY OF IMAGE PROCESSING FOR TRAFFIC DATA COLLECTION

In this section, a number of potential applications for video image processing in the traffic data collection and surveillance field will be examined and the present capabilities of image processing systems, together with foreseeable developments, relevant to each application, will be discussed.

Measurement of the number, speed and length of vehicles as they pass a point on the road network

Moving vehicles passing through a scene can be detected by measuring a change in the grey-levels of individual pixels or groups of pixels. This can be thought of as projecting simple light sensors onto the carriageway and can be achieved by setting up a camera in an overhead position looking down on the roadway. Two windows could then be superimposed across the image of each traffic lane and the scene sampled at high frame rates (up to 200 frames/s). The time at which each vehicle is observed/detected at a window would later be compared with the time of arrival/detection at the next window and thus an estimate of speed, vehicle length and lane occupancy could be derived if the spatial separation of the road elements corresponding to the two window positions is known.

Before applying the above techniques, a number of problems associated with background frame differencing have to be overcome and it is worth discussing the most relevant of these; that of choosing and updating the reference frame. Even on an apparently dull day a variation in cloud thickness can cause a quadrupling of the ambient light level within seconds. As noted earlier, it is usual for the camera's automatic-gain-control or automatic-iris to respond to the total scene brightness, therefore scene content, including vehicles passing through the scene, can affect the apparent luminance of the background. Driving an auto-iris from a photodiode looking at an inactive part of the scene is possible, but in practice it may be difficult to match the photodiode response to that of the camera. Alternatively, the background frame can be updated by monitoring selected pixels from inactive parts of the image. This, however, would require operator intervention to set up the system on site and could be difficult to implement in a scene with heavy shadows from nearby buildings etc.

A more attractive idea is to apply a double thresholding technique to each difference image, an upper one to isolate all valid movement points and a lower one to detect small changes which will be assumed to have been caused by slight changes in ambient light. Thus small changes could be used continuously to update the background, whilst significant changes would not affect it. This approach has been used by the authors in the development of their image processing system described in a later section of this chapter.

Automatic surveillance of traffic throughout a 1-km length of roadway

While attempts to monitor short stretches of highway are considered quite feasible, the question of whether a system can be built which will track a multitude of individual vehicles within a wide-area scene such as a 1-km length of motorway is more difficult and involves many other factors. It seems highly improbable that simple detection lines or windows would be capable of such a complicated task. The success of simple systems relies on constraining the processed image. For more general wide-area detection, such systems would fail due to a lack of information with which to describe the likely range of events. With a windowing system, it is unlikely that, as a vehicle passes through the predefined group of pixels, a satisfactory description of the vehicle can be obtained and, consequently, it is doubtful that any description of this kind could be correlated with similar data from a sequence of images to derive individual vehicle movements. Under these circumstances, therefore, a more sophistocated system will be needed. Tracking systems, using the cross-correlation search and feature matching techniques, have been tried and it has been possible, although not in real-time, to analyse a sequence of images and estimate vehicle speed (Schlutsmeyer, 1982). However, three important factors related to the achievement of this objective are occlusion, image resolution, and processing time. These points are discussed briefly in turn below.

The first problem is that of occlusion or occultation, where one object obstructs the view of another object. With an oblique view of a traffic scene, especially if a high proportion of heavy goods vehicles are present, even grey-value signatures would prove difficult to track, due to change of shape and grey-value caused by partial occlusion of any vehicle travelling in a platoon as it approached the camera position. The most successful approach is likely to involve detection of a vehicle in the foreground, while it is free from occlusion, and subsequent tracking of the receding vehicle.

Secondly with regard to image resolution, Figure 3 shows the number of pixels occupied by a vehicle against distance from the camera, for various image resolutions. The plot assumes a camera with a 60 degree field of view, mounted at a height of 22.5m and observing a level highway between 10m and 1km from the observation point. If, for a high probability of detection, a vehicle showing an area of 5 square metres must be present in a 3 x 3 matrix of pixels, then Figure 3 indicates that to observe vehicles over 800m of roadway would necessitate using a camera with a resolution of 1024 x 1024 pixels. The majority of cameras in use for motorway surveillance today have a resolution of about 256 x 256 pixels and therefore have a range of about 200m on the basis of Figure 3.

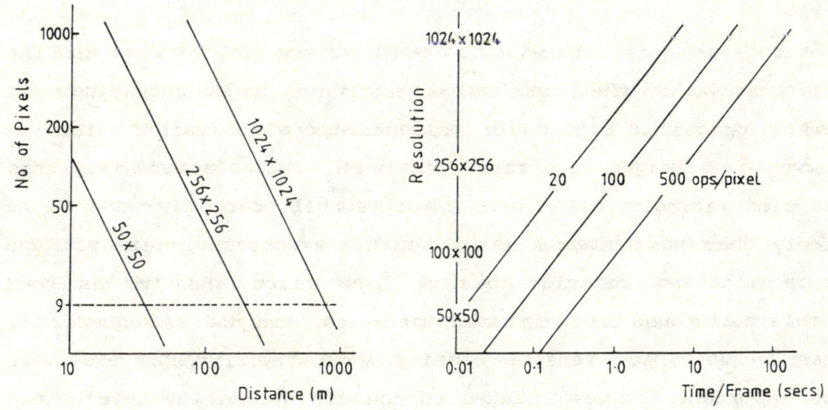

Figure 3 (left). Distance from camera versus number of pixels occupied by a vehicle for several image resolutions (60^0 field of view and vehicle of area $5m^2$).

Figure 4 (right). Time to analyse scenes of various resolutions (executing five million instructions per second).

Although traffic further away could be observed if a zoom lens were fitted, closer traffic would then be lost.

Thirdly, the computational requirements to analyse scenes of various resolutions can be seen from Figure 4. With a resolution of 1024 x 1024 pixels, it can be deduced that it would take a computer, executing five million instructions per second, in excess of 20 seconds to perform a relatively straightforward algorithm, consisting of 100 operations per pixel. Since it is desirable to track many vehicles in each frame the processing time could be significantly longer. Research carried out by the Jet Propulsion Laboratory (JPL) for the U.S. Federal Highway Administration confirms that wide-area surveillance using correlation search methods is a computationally expensive procedure. It was suggested that their system would be able to track only three vehicles at once when fully developed (Schlutsmeyer, 1982).

Vehicle tracking through a junction

Accepting that difficulties exist with surveillance, it is thought likely that within the next five to ten years image processing will become a practical method for collection and analysis of turning movement information at road junctions. In general, systems capable of tracking individual vehicles will be complex and it is unlikely that real-time multiple vehicle tracking systems will be developed, at an acceptable cost, for some years. Systems designed to analyse video-tape images, off-line, should be available earlier. Many problems associated with the interpretation of cluttered street scenes will be encountered during the development of a tracking system for turning movement counts. However, occlusion can be minimised by near-overhead camera positioning and an image resolution of about 100 x 100 pixels should be adequate, whilst also ensuring that processing time is not excessive.

Hardware is now available which can grab and digitise video images from a VTR at 25 frames/s. With a 'frame-grabber' it would be possible to develop an off-line system which could access the same video recording several times, and with each pass carry out a portion of the total processing task. In this way it would be possible to count and determine the direction of movement of vehicles as they pass through a junction, by considering one entry or exit at a time.

The possibilities for real-time processing of the complete image of a simple junction are thought to be good, but the development of robust algorithms, with which to extract data from the traffic scene, would have to be followed by their implementation in special-purpose hardware if the desired processing speed is to be achieved. The current lack of suitable hardware should not be of particular concern, since a range of alternative hardware options will be available once the processing steps have been determined (Fountain, 1981 and Reeves, 1984).

It is considered that for junctions covering an area of say 30m square, where occlusion is not serious, reference frame differencing and thresholding may prove the simplest yet most effective solution. Once the vehicle has been identified, then its centroid or leading edge can be tracked from frame to frame. By noting speed and direction from past frames, it is possible to perform a heuristic prediction and forecast the location of the vehicle in the next frame. With a sufficiently powerful algorithm, this method would allow occluding movements also to be analysed.

Vehicle classification

The ability of a vision system to classify vehicles generally as 'CAR', 'BUS' or 'HGV' depends upon the availability of practical definitions that distinguish all vehicles that are cars, buses or HGVs from all those that are not. If some form of template is used to fit the vehicle to a classification code not only must the template describe the class of vehicle, but the image must be capable of segmentation into shapes which might reasonably be assumed to be vehicles, so that these can then be tested for classification.

Template matching systems traditionally operate by storing two-dimensional representations of a range of reference images which are compared on a pixel-by-pixel basis with the image of each object to be classified or tested. In most practical applications of template systems, e.g. in non-contact inspection or parts sorting and assembly, the image of an object can be captured when it is a known distance from the sensor and the expected variation between the reference image and the test image is small. In the traffic application this would not be the case. Many different reference images could be produced for a given class of vehicle, but it seems unlikely that a single two-dimensional silhouette of a particular class of vehicle could be satisfactorily used to classify the likely range of shapes and sizes contained in each vehicle group.

A more promising approach would rely on the extraction of those features, e.g. overall vehicle length and height, or the number and position of wheels, which best describe different classes of vehicle. It seems feasible that if vehicles were to pass a side-mounted camera in single file, then the features described above could be measured. A system which first detected the presence of a vehicle in the centre of the scene by scanning every image, then reverted to a slower but more detailed analysis of the vehicle image, would appear to be the most appropriate. By applying sufficient constraints, the distance of the vehicle from the camera could be kept within acceptable limits even if lateral position caused some uncertainty.

In the longer term, if the classification of vehicles in a wide-area scene is still desirable, much more powerful hardware and complex software will have to be developed. Three-dimensional models of each vehicle type would have to be held within the computer as reference vehicles and feature matching undertaken with the acquired images as vehicles cross a reference line in the scene. Whilst this approach has been used successfully off-line to detect buses (Hogg et al, 1984), it would be far more difficult to detect other vehicle types on account of their wide variation of possible shapes and sizes.

(3) REVIEW OF IMAGE PROCESSING SYSTEMS USED FOR TRAFFIC DATA COLLECTION

Compared with the total research effort in the field of computer vision and image processing, that directed towards traffic data collection and monitoring has been very limited. However, during the past decade a number of image processing systems have been developed with traffic data collection in mind, the first being Japanese systems dating back to the early 1970's. Since that time, further development work has been undertaken in Japan and broadly similar systems have been developed in several European countries, notably in Belgium, Sweden and the United Kingdom. Over the same period the U.S. Federal Highway Administration has been funding research into image processing applied to freeway surveillance and the Australian Road Research Board has recently developed a hardware-based video detector system for traffic counting purposes.

A brief description of some of these current systems follows, making a distinction between systems designed to collect traffic data at a point on the road network and those intended to monitor or track vehicles within a wide area.

Point detection systems

Since the late 1970's the authors have been concerned with the development of a portable traffic data acquisition system based on image processing technology. The initial system, which was largely hardware based, comprised a CCD camera interfaced to a microprocessor pipeline. A magnetic tape cartridge unit was used for storing the recorded data for subsequent off-line analysis. Field trials carried out with this system illustrated the problem of changing ambient light levels and the need for greater flexibility in capturing and updating the stored reference frame ; nevertheless some 88% of vehicles passing through the scene during a typical 30-min. observation period were automatically counted (Dickinson et al, 1983).

More recent developments have resulted in the TRIP (Traffic Research Image Processing) System which is a flexible development system utilising an Intel 86/330 computer and a Reticon photodiode camera with resolution 100 x 100 pixels. Short sequences of recorded traffic data are stored on disk so that different software algorithms relating to reference frame updating and thresholding routines can be tested with the same data set. To reduce the amount of data to be processed from each frame, a 'windowing' system is used, with up to ten windows of various shapes and sizes being defined by the operator or recalled from memory. These windows are analogous to loop detectors for speed and occupancy measurement purposes and a typical layout for a two-lane single carriageway is shown in Figure 5. This has allowed data to be collected at a rate of 60 frames/s, which is necessary to achieve an acceptable order of accuracy for individual vehicle speeds.

Figure 5. Use of the windowing technique for speed and occupancy measurements on a two-lane carriageway.

Results based on a sample size of 974 vehicles observed in various weather and lighting conditions have shown that over 99% of vehicles were detected by the TRIP system, with standard errors of 7%, 11% and 9% for vehicle speed, occupancy and length respectively (Dickinson et al, 1985). A limited number of observations during hours of darkness have indicated that it should be possible to count vehicles and measure their speed at night by observing vehicle headlights and taillights. It is intended that the various software algorithms developed and tested with the TRIP system will eventually be committed to hardware, thus allowing the image processing system to operate in real-time, either for traffic data collection purposes or for traffic monitoring within the framework (4) of an Automatic Incident Detection (AID) system.

Somewhat similar image processing systems, but using special-purpose hardware, have been developed by Takaba et al (1984) and by Maes et al (1984). In both these systems, window data have been used so as to allow operating speeds of 50 frames/s. Field trials have produced vehicle detection rates in excess of 90%, but problems relating to sensor blooming and the inability of the camera's auto-iris control to adjust to rapid variations in light intensity have been encountered.

Another windowing system developed in Sweden has been described by Abramczuk (1984). Receding traffic is viewed along the roadway from an elevated camera position within a 200m long detection zone of one pixel width. Real-time, on-line tests of the system have not yet been performed, but some simulation runs using a digitised video film sampled at a rate of approximately 4 frames/s have been undertaken.

The hardware-based system developed by the Australian Road Research Board (Dods, 1984) uses a standard vidicon camera and processes data from up to 16 detection points which can be positioned by the operator anywhere within the video frame. The performance of the system using a pre-recorded videotape taken on a bright sunny day showed a vehicle detection rate of over 97%.

Area detection systems

In contrast with the point detection systems described above, the Federal Highway Administration, U.S.A., has financed research at the Jet Propulsion Laboratory, Pasadena, into the development of a Wide Area Detection System (WADS). The work was initiated in 1976 and is aimed at tracking vehicles within a traffic scene, typically along a length of freeway. By observing receding traffic from an overbridge, a vehicle is detected as it crosses a line in the foreground of the image and its 'signature' is obtained. Cross-correlation search techniques are then used to track the vehicle through successive frames, as previously explained, and from the resulting information traffic stream parameters can be estimated.

Apart from the search procedure being computational expensive, difficulties have been encountered due to shadows from roadside objects or adjacent vehicles affecting the grey level signatures. Also a high resolution system is required to track individual vehicles at distances remote from the camera location so that a maximum practical range of about 120m has been suggested. Saxton (1984) has described freeway trials of WADS which resulted in 49 missed vehicles or overcounts in a sample of 1085 vehicles and has cited problems relating to vehicle shadows, reflections from vehicle windshields or brightwork and the effects of rain causing water spray; however, the performance under good visual conditions was felt to be similar to that of loop detectors.

A vehicle registration number recognition system

For completely different purposes than those described above, the British Home Office has recently developed and tested an image processing system capable of automatically reading vehicle registration numbers. The system consists of a camera linked to specially-developed high-speed image processing hardware and then via telephone lines to a data bank which contains the registration numbers of all recently-stolen vehicles or non-valid numbers.

When a vehicle is a pre-set distance away, a video frame is stored, the position of the vehicle number plate is determined and the latter is then analysed by comparing with standard alphanumeric templates held within the system. When a full registration number has been estimated, it is checked with the central data bank and if the vehicle has been reported stolen, the nearest police car can be automatically alerted. The whole process takes little more than a few seconds. The system was tested from an overbridge on the M1 motorway, but although technically successful, the equipment is considered to be too expensive for normal usage.

(5)

CONCLUSIONS

From the above, it can be seen that a number of image processing systems designed for specific traffic monitoring applications are currently under development in various countries throughout the world. However, despite recent advances in microprocessor technology, all existing image processing systems have been forced to accept a compromise between the use of a high resolution image and the desirability of operating in real-time with a frame rate of at least 50 frames/s. Whilst this dilemma can be overcome, to some extent, by confining the analysis to small areas of the overall scene, this is not possible where a wide-area traffic monitoring system is required. Of the four specific objectives discussed earlier, that relating to the collection of traffic data at a point on the road network appears to be the most easily attainable and has been achieved by a number of investigators. Difficulties still exist, however, in relation to rapidly changing ambient light levels, the shadowing effects of vehicles, night-time observations and the need for an elevated yet stable camera mounting. Technical problems such as sensor blooming can also cause difficulties, although these can be overcome to a considerable extent by using the more expensive solid-state cameras instead of standard vidicon cameras. Wide-area surveillance is more difficult on account of the need for high resolution systems capable of analysis in real-time at relatively high frame rates. The problem of occlusion is also inevitable with such systems if individual vehicles are to be monitored, thus limiting the effective distance over which the system could operate in practice to a maximum of about 200m.

The most suitable arrangement for using image processing within the context of an automatic incident detection system appears to be one in which television cameras are allowed to operate in a dual mode. Each camera could automatically monitor traffic flow parameters within a short length of roadway close to the camera position and give an alarm signal in the control centre when speed or occupancy reached pre-determined critical levels. This would alert control room staff who would verify manually from the television picture the authenticity of the generated alarm signal and then use the camera's pan, tilt and zoom capability to identify the nature of the incident and notify the appropriate emergency services. Such an arrangement would eliminate the need to provide both loop detectors and TV cameras in traffic surveillance systems, as is common practice at present.

Although tracking of individual vehicles through a junction suggests similar problems to those mentioned above for wide-area surveillance, the possibility of using repeated off-line analysis of recorded videotapes to build up a record of turning movements offers greater promise and should be achievable within the next decade. Thereafter, on-line analysis of vehicle movements through a simple junction might become possible with enhanced computer systems incorporating parallel processing capabilities.

The final objective, vehicle classification, appears much more difficult, other than using a crude classification according to vehicle length. Whilst it has already been shown possible to recognise a unique type of vehicle such as a bus using a three-dimensional computer model, the more general application of such methods to identify cars and commercial vehicles which have a wide variety of shapes and sizes appears remote. Indeed, it seems unlikely that image processing systems will ever approach the capabilities of human vision in its ability to interpret complex traffic situations; they are, however, expected to find increasing application in undertaking the more routine tasks in the traffic monitoring and data collection areas over the next ten to fifteen years.

REFERENCES

Abramczuk, T. (1984). A microcomputer based TV-detector for road traffic. Seminar on Microelectronics for Road and Traffic Management. Traffic Bureau, National Police Agency of Japan, Tokyo, pp.87-96.

Dickinson, K.W. et al (1983). Traffic data collection and analysis using video image processing. Proc. Seminar on Traffic Operation and Management. PTRC, London, pp.95-106.

Dickinson, K.W. et al (1985). Road traffic data collection using the TRIP system. Proc. Seminar on Transportation Planning Methods. PTRC, London, pp.135-146.

Dods, J.S. (1984). The Australian Road Research Board video based vehicle presence detector. Proc. Internat. Conf. on Road Traffic Data Collection. Inst. Elect. Engrs, London, pp.96-100.

Fountain, T.J. (1981). CLIP 4: a progress report. In: Languages and Architectures for Image Processing, Duff, M.J.B. and Levialdi, S. (Ed.). Academic Press, London, pp.283-291.

Hogg, D.C. et al (1984). Recognition of vehicles in traffic scenes using geometric models. Proc. Internat. Conf. on Road Traffic Data Collection. Inst. Elect. Engrs, London, pp.115-119.

Maes, W. et al. (1984). Development of a detection system based on the use of a CCD camera. Seminar on Microelectronics for Road and Traffic Management. Traffic Bureau, National Police Agency of Japan, Tokyo, pp.78-86.

Mortelmans, J. and Venstermans, L. (1984). Analysis of video recordings and data processing. Proc. Internat. Conf. on Road Traffic Data Collection. Inst. Elect. Engrs. London, pp.101-104.

Nagel, H.H. (1983). Overview on image sequence analysis. In: Image Sequence Processing and Dynamic Scene Analysis, Huang, T.S. (Ed.). Springer Verlag, Berlin, pp.19-213

Reeves, A.P. (1984). Parallel computer architectures for image processing. Computer Vision, Graphics and Image Processing, 25(1), pp.68-88.

Saxton, L. (1984). State of the art in highway electronics systems - two examples. Seminar on Microelectronics for Road and Traffic Management. Traffic Bureau, National Police Agency of Japan, Tokyo, pp.39-54.

Schlutsmeyer, A.P. (1982). Wide area detection system (WADS). Report No. FHWA/RD-82-144. Federal Highway Administration, Washington,DC.

Takaba, S. et al (1984). A traffic flow measuring system using a solid-state image sensor. Proc. Internat. Conf. on Road Traffic Data Collection. Inst. Elect. Engrs. London, pp.110-114.

Wootton, H.J. and Potter, R.J. (1981). Video recorders, microcomputers and new survey techniques. Traff. Engng Control, 22(4), pp.213-215.

EDITORS' NOTES

(1) There are a wide range of technologies suitable for the continuous collection of traffic data. Chapter 2 describes not only inductive loops but also systems using infra-red, ultrasonic, microwave and other media.

(2) The advent of hand-held data loggers goes some way to overcome transcription problems inherent in traditional manual techniques of data collection but does not reduce the manpower requirement at the roadside.

(3) Chapter 2 describes how vehicles may be classified using appropriate combinations of more conventional vehicle sensors.

(4) Another approach to automatic incident detection, mentioned in Chapter 6, makes use of data from inductive loops.

(5) As is mentioned in Chapter 3, a 70 percent accuracy has been achieved with this system.

Information Technology Applications in Transport, pp. 87–104
P. Bonsall, M. Bell (Editors)

Chapter 5

ELECTRONIC TICKETING SYSTEMS

Andrew Mellor[1] and Peter White[2]
[1]*Halcrow Fox Associates, UK,* [2]*Polytechnic of Central London, UK*

INTRODUCTION

This chapter examines the recent development, and future potential, of electronic systems for collection of bus fares and car parking fees. The two issues are related, not only through an application of similar technology (often produced by the same manufacturers), but also serving, broadly speaking, the same market - local transport.

The greater part comprises an examination of the electronic ticketing systems (ETS) now used in the bus industry, with a shorter comparative review of parallel developments in car parking systems. Our coverage of ETS in the bus industry is based on, and updates, our recent study for Science and Engineering Research Council (reported on by Mellor, 1985). Our aim in that study, as in this chapter, was to consider the application of the technology rather than the technology itself, since technical details would become out of date almost as soon as they are published.

ELECTRONIC TICKETING SYSTEMS IN THE BUS INDUSTRY

Data Requirements

There is a growing demand for better management information in the bus industry. In an increasingly competitive environment there is a need to monitor the performance of services and develop an informed response to fluctuations in demand or changes in market conditions. However, data collection and analysis has, in the past, been an expensive and time-consuming business.

Surveys to obtain the necessary data have been mounted on an ad-hoc basis by most operators. Few have attempted a regular programme, and almost all rely on small samples from each route covering a very limited time period. Yet every time passengers board a bus and pay their fares they are effectively interviewed about their journeys. Traditional mechanical ticket machines can only retain a fraction of the resulting information - usually only a summary of the total revenue and count of fare-paying passengers for each driver shift or day. Even the latter is of uncertain accuracy if multiple ticket issue is required for individual journeys. These problems have been particularly significant in countries like the UK where graduated fare scales have traditionally predominated.

Development of ETS

The development of electronic ticket machines since the late 1970s has provided the means to capture details of every transaction made on the bus and appears to offer exciting opportunities as a low cost method of continuous monitoring. The concept of an electronic ticketing system is shown diagramatically in Figure 1. The computer and software for analysis of the data stored by the ticket machine should be viewed as an integral part of the system and not mere appendages to it.

The first example of the new technology to be introduced in Britain (or elsewhere, to the authors' knowledge) was the 'Timtronic', developed by Ticket Equipment of Cirencester (now Almex Systems Ltd), prototypes of which entered service with the National Bus company subsidiary Crosville, in North Wales in July 1978. They were followed by Control Systems Ltd (formerly Bell Punch) who produced the 'Autofare 3', an electronic variant of their farebox system with remote ticket issue, and subsequently the self-contained 'Farespeed' to meet other applications. A company new to the ticketing field, Microsystem Design Ltd., entered the field with the

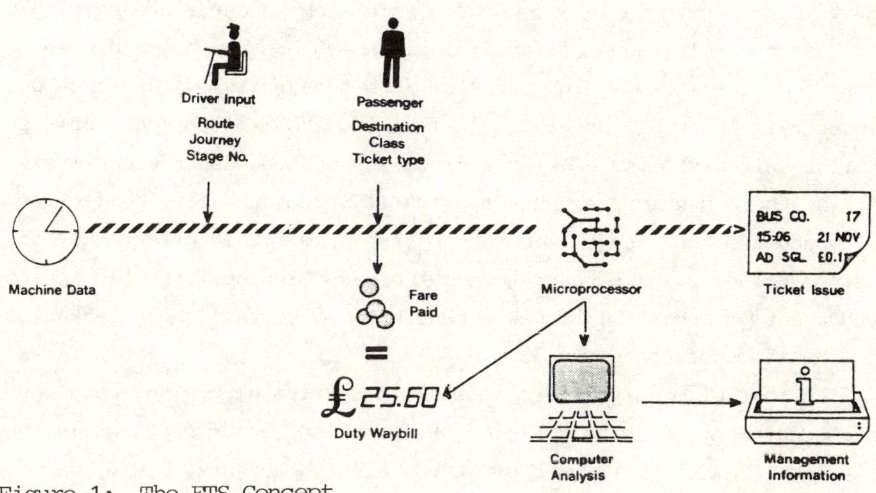

Figure 1: The ETS Concept

	MECHANICAL TICKET MACHINES	ELECTRONIC TICKET MACHINES		ON-BUS SURVEYS	
		Programmable Fares	Fare Table Look-up	Census	Interviews/ Questionnaires
REVENUE					
By Route	✓	✓	✓	✓	✓
By Ticket Type	d	✓	✓	✓	✓
By Passenger Type	X	✓	✓	✓	✓
PASSENGERS					
Fare Paying (on-bus)	d	✓	✓	✓	✓
On Travelcards/Seasons	X	p	p	✓	✓
On Concessionary Permits	X	p	p	✓	✓
By Ticket Class	d	✓	✓	✓	✓
By Route	d	✓	✓	✓	✓
By Time of Day	l	✓	✓	✓	✓
Boarding by Stage	X	✓	✓	✓	✓
Alighting by Stage	X	X	✓	✓	✓
Origin-Destination	X	X	i	i	✓
Linked Trips	X	X	X	X	✓
Journey Purpose	X	X	X	X	✓
Personal Characteristics	X	X	X	l	✓

✓ = yes d = dependent on system p = possible (use depends on conditions)
X = no i = incomplete/partial l = limited

Figure 2: Comparison of Data Collection Potential

'Wayfarer', originally a low-cost basic machine but since developed to offer a full range of facilities, which has now been adopted by many UK operators. In parallel with this activity in the bus industry, Thorn-EMI developed 'APTIS' and 'PORTIS' to meet British Rail requirements for booking office, and on-train use respectively. All of these machines are based primarily upon cash fare collection, although pass use can be recorded by the driver. In effect they are built around existing operating methods and, in the first two cases, represent a shift into new technology by established ticketing equipment manufacturers.

A number of manufacturers in other countries have also developed such equipment. Perhaps the closest equivalent to UK practice is the 'S-Light' machine produced by Sanyo Electronic Works Ltd of Japan, which has been supplied in large numbers to Singapore Bus Services Ltd, based on a farebox/remote ticket issue layout similar to 'Autofare'. Camp-CGA of France, Sadamel of Switzerland, and the Australian firm AES have also produced on-bus ticketing machines. The number of firms involved in the production of ETS equipment is now growing steadily.

A major obstacle facing the equipment designers was the harsh in-vehicle environment where the machines are subject to vibration, dirt, damp and electrical interference. Now initial teething troubles have been overcome the machines peform well and many are expected to have a life of ten years or more.

The extent of the potential revolution in data collection is demonstrated in Figure 2 which sets out the capabilities of mechanical and electronic ticket machines, with the data available through on-bus surveys included for comparison.

Electronic ticket machines are now rapidly replacing their mechanical predecessors in Britain, offering advantages in terms of revenue accountability, security and reliability. Capital costs for re-equipping a fleet of buses are now comparable and moving increasingly in favour of electronic systems, since only one machine per bus is required, compared with one per driver for older systems.

The question is no longer whether electronic machines should be adopted but rather what use will be made of the information they can provide.

Reaction from crews who use the systems has been extremely favourable, most being quick to learn the new techniques required despite some initial scepticism.

Standardisation

The absence of an industry wide standard specification for ticket equipment is clearly a problem. In the UK the National Bus Company did considerable work in the mid-1970s on a specification which formed the basis for the Timtronic design, but not all the subsidiary companies thought this approach to be desirable. The Passenger Transport Executives (PTEs) who run services in seven major conurbations, and London Transport, set up a 'Through Ticketing Working Party' with a view to producing a set of common standards. A team of design consultants was engaged to look at the problem, reconcile differences and produce a working model. However the proposed concept proved costly, and although some ideas may be taken up, most PTEs have now gone their separate ways. The result has been a proliferation of machine designs to suit individual applications; which represents a major source of frustration to the suppliers and ultimately greater cost to their customers. A similar lack of standardisation has been evident in the definitions employed for statistics collected in the industry. No effective overall co-ordination has been applied by either the UK bus industry's trade association (The Bus and Coach Council), or the Department of Transport.

The Decision to Invest in ETS

A recent survey, carried out by the authors among a wide-ranging sample of UK bus operating companies, indicated that the following priorities were important in their choice of system:

* All sectors of the industry were united in rating speed and ease of operation as their principal consideration in the choice of a system. This factor was rated as first priority by 75 percent of

the operators surveyed and first or second by no less than 92 percent.

* Data capture facilities were high-rated by two-thirds of operators.

Operators as a whole saw more detailed management information as the most useful benefit likely to accrue from the introduction ETS but among undertakings with a high level of urban operations, correct ticket issue was seen as the principal gain. Other benefits cited included faster availability of management information, improved cash accountability and potential for reducing administrative staff time. Improved computer awareness and computer literacy were also seen as useful spin-offs.

* It was found that the views of operators on the usefulness of ETS in providing management information were frequently coloured by the extent of information currently available to them. This can vary quite markedly. For example, in performance monitoring most operators allocate revenues to routes and attempt some form of cost allocation; occasionally involving quite sophisticated analysis by time of day. A few, however, still get by comparing overall income and total costs without any clear objective base for decision making. Where information is at present limited in scope, expectations appeared to be lower. The broadest horizons were to be found among operators already using ETS successfully.

The availability of software for data analysis was not accorded a high priority by many operators. This is interesting,given that adoption of a system with inadequately developed software has been the cause of the most serious and lasting dissatisfaction with ETS to date.

Many operators do not yet seem to have "thought through" the implications of introducing electronic systems. Traditional working practices and conventions may no longer be appropriate with the new technology and simply automating poor practice will inevitably lead to disappointing results. Manufacturers have a responsibility here. It cannot be acceptable for manufacturers to claim they only supply what the customer asked for; they must take a lead in educating operators as to both the potential of their systems and, equally as important, the limitations.

Small operators may be understandably unwilling to invest substantially in new technology, especially if their stage carriage activities are seen as marginal to their coaching work. However, they too could benefit from improved data. One major constraint for a small fleet may not be the cost of the on-bus machines themselves, so much as the hardware and software needed for analysis at the office, where the cost has to be spread over ten vehicles rather than a hundred. Use of existing business micro-computers, rather than specialised hardware, is the key. The ACT Apricot and IBM PC are already popular for data analysis and where this can be combined with word-processing or coach booking systems, the economics look very favourable. Independent operators are already making extensive use of specialised programmes for coach tour and private hire bookings (vide recent article in the Coaching Journal - 1985). Reduced cost of the hardware to read data modules could obviously be critical. (1)

Software

The collection of data is always a means to an end and not an end in itself. This may sound a truism but it is alarming how little attention is paid to the fact that the quality of management information derived from ticket machine data can only be as good as the software used for its analysis. Unfortunately, thus far, advances in hardware (the ticket machines themselves) have not been matched by software development. In many cases, it is almost completely absent.

There have been a number of contributory factors to this situation but a major one has undoubtedly been the lack of clarity or consistency in operators' stated requirements. Software development is expensive and the ticket machine manufacturers have little incentive to produce more than the minimum required to sell their product. As a result the majority of operators with ETS use data analysis packages which produce only basic statistical reports.

Even where packages have been developed, dissemination of their application has been very slow. For example, a package to analyse data for the 'Timtronic' machines introduced by Eastern Scottish (a subsidiary of the Scottish Bus Group) was produced by Expert Systems Ltd, but does not appear to have been applied outside SBG, although some of its features appeared in a package from Almex over three years later.

Applications of ETS data

It was clear from our survey that electronic ticket machines will become the primary means of data collection for management information purposes, eliminating the need for many ad-hoc surveys. The availability of a wide range of additional information on a continuous basis for the first time opens up new areas for investigation, many of which will be of increasing importance in a deregulated industry. For example:

Marketing The ability to study travel patterns of particular groups of passengers such as the elderly, or use of various ticket types by time of day down to route level offers the opportunity to revolutionise the marketing of bus services.

Monitoring Demand Disaggregate elasticity measures may be obtained through observing the effects of service level or fare changes on patronage at different times, or for given journey lengths. Since calculations are based on a continuous flow of data representing a large sample of passengers, the results can be used with greater confidence than those based on a few isolated instances in the past. Continuous monitoring also allows the accuracy of the predictions to be assessed. This feedback will be invaluable when considering future changes.

Negotiation of compensation payments for concessionary fares ETS can log all journeys made using concessionary permits (assuming a token fare is payable or pass use is recorded) and can provide detailed information on passenger demand. This data should give operators a solid base for negotiation of compensation from local authorities.

Tendered Services Under the Transport Act, 1985, general network revenue support will no longer be permitted in Britain.

Competitive tenders must be sought for specific 'non commercial' services which local authorities wish to retain. On these, the authority may specify the data to be collected, and the type of machine to be used. Services are already operated under tender on behalf of London Regional Transport with such requirements.

Limitations Electronic systems represent such a significant advance on our previous ability to collect data that it is sometimes easy to forget there remain several significant limitations. Computer analysis of data may give a spurious impression of precision and common sense is a vital asset in interpreting the results. (2)

Perhaps the most serious limitation concerns the recording of passengers who do not need to pay a cash fare on entering the vehicle, such as holders of travelcards, scholars' passes, or concessionary permits for free travel. Among large city operators in particular, the use of such systems has been encouraged both through local authority policies (especially in concessions), and by managers as a marketing device. In some cases over 30 percent of passenger trips are now made by holders of network tickets, with a further 20 percent on concessionary passes. For further information see White (1984). With present generation of electronic ticket machines one is reliant on the willingness of the driver to record usage, by entering details into the machine when a passenger boards. In the case of some low-density routes it may be practicable to record passenger type, stage boarded and intended alighting point by interrogating the passenger as for a cash fare. However, on high-density routes this is clearly impractical; the extra time involved would negate one of the major benefits of introducing such tickets. At best a simple boarding count might be obtained. In such situations supplementary data collection is required, and several major operators prefer to rely on such surveys for all their planning data.

Bus passenger surveys When the survey sample size is large, as it is when calculating overall network characteristics, there are virtually no advantages in using the partial data provided by the ticketing system over more detailed surveys. In these instances the degree of confidence associated with the survey results is high. However as the area under scrutiny diminishes with a consequent reduction in the sample the errors for the surveyed data become proportionately greater and the level of confidence drops. At route level the data produced by ETS- a 100 percent sample of cash fare paying passengers - begins to look more attractive. Our analysis of continuous on-bus survey data (from West Midlands PTE) showed that, at route level, a more reliable estimate of overall passenger journey characteristics (such as average journey length), may be produced by ETS than by a single sample survey irrespective of the level of off-bus sales.

Data from continuous on-bus surveys by both the West Midlands and Greater Manchester Passenger Transport Executives was examined, to determine to what extent data for cash ticket sales recorded through ETS could be regarded as a suitable proxy for overall characteristics. In the instance of average passenger trip length, this was found to be the case. Where market penetration of travel-cards is low (usually due to a relatively high price) their average trip length is significantly higher than for cash fare users (since, on graduated fare scales, savings from travelcard use are greater for the longer trips). However, the effect on the overall average weighted mean trip length is small. Conversely, where market penetration is high, there is little difference in average trip length. Generally speaking, the mean trip length for cash fare users is within ± 1 standard error of that for all passengers, at the 95% confidence level.

Cash fare ETS data may thus be used to estimate mean trip length, and hence total passenger-km on a network (provided that the number of non-cash trips is known). However, if a distribution of all passenger trips by time of day was sought, cash fare data could give a very misleading picture of the market over the day as a whole, as travelcard penetration is much higher at peak periods.

One clear advantage that data from ticket machines has over surveys is the continuous nature of data collection which eliminates the uncertainties associated with daily and seasonal variations.

Knowledge of the degree of variation, provided by ETS, can greatly assist in the planning of surveys to provide a more representative sample, and in interpretation of the data thus obtained.

Overall it is better to consider ticketing systems and surveys as mutually beneficial data collection techniques rather than as direct alternatives. Surveys can provide data on items such as linked trips and actual origin and destination and as such can never be fully replaced. There are however sound economic reasons for reducing the extent of surveys and the results from ETS can be used to direct survey resources in a more efficient manner.

Future developments

Software We have already stressed the need for software development to match the hardware already available in order to produce management data. This will enable the present range of machines to play their full role.

There is, in our view, a desperate need for a more versatile software package to input to, and to be integrated with, other administrative tasks such as wages and mileage. Facilities for exception reporting - flagging results outwith a defined tolerance band - are especially important as this prevents management being overwhelmed by the sheer volume of data.

Data analysis would, ideally, be able to be carried out at various levels of disaggregation; geographically (to the complete network or for individual route journeys or even journey stages) and by market segment (ie ticket type, class of passenger or time of day). Analysis could be carried out for reporting periods specified by the user.

The problem of cash handling Software advances notwithstanding a major limitation remains in that cash handling is still the main determinant of boarding times. Electronic machines used in Britain to date have done nothing new to alleviate this. Two surveys conducted as part of our study illustrate this point. A sample of 1060 passenger boardings in the West Midlands gave an average of 2.5

seconds for 'Autofare 3' users (ie paying cash with exact fare, no-change policy), and 2.2 seconds for pass holders. In contrast, Timtronic machines in Oxford, employing automatic fares look-up (from input destination stage), but with manual change-giving, produced an average of no less than 6.4 seconds per passenger, compared with 2.6 seconds for pass holders in that city (based on a total sample of 554). Change-giving thus emerges as the dominant element in boarding times. In some cases, notably the earlier Wayfarer machines, slow speed of ticket printing may also be a problem, but where change giving applies, the driver can do this while the ticket is being printed.

Use of passes eliminates the delay caused by cash handling, but may be appropriate only for certain types of travellers such as frequent and regular bus users. An exact fare, no-change policy may be inconvenient for passengers and still involves large scale coin counting and handling by the operator. Less frequent users are unlikely to buy travelcards, except at very low prices, and further extension of concessionary free travel is unlikely given the constraints on local authority spending. There is in any case the problem of recording the use of such passes.

Magnetically-encoded tickets A possible solution to these problems appears to lie in the developments which have recently been taking place in the area of magnetically-encoded tickets. These have been around for some time on urban rail systems, initially for single tickets and more recently for travelcards (as in Tyne and Wear) and stored value tickets (as in Hong Kong and Washington DC). Developing technology has produced more compact devices, making readers a practical possibility on buses for the first time. The Westinghouse Cubic 'Fast Fare' is one such machine currently in use in Long Beach, California. Similar machines from other manufacturers including Camp and Sanyo have been undergoing trials in Hong Kong and Singapore. A ticket may be slid quickly through a "swipe reader" for checking and, if necessary, re-encoding. No cash transactions are required on the bus.

If this type of equipment was fitted at bus entries, travelcards and similar passes could be read, or a flat fare deleted from a stored value ticket. If graduated fares applied, exit reading would also be necessary. Although pay-on-exit systems are in regular use in Japan the operation of such a system, in conjunction with the more normal pay-on-entry for cash fares, could produce an undesirable increase in stop times ; nevertheless this type of system was under consideration in Singapore. The maximum fare would be deducted on entry and the unused portion credited on exit to ensure compliance.

A more advanced system is under development by Eureka Systems of Slough UK. Their approach is to use an encoded tag which electronically transmits stored data to an on-bus sensor without need for visual display or physical contact with a reader. On entering the vehicle, the passenger carrying such a tag would be registered and their class identified to the driver. Automatic re-encoding on leaving the vehicle would also be possible, allowing passengers to be charged for journeys made The 'tag' - already used in some security applications - measures only a few centimetres across, and may be carried in the pocket or bag of the holder. It has a life of about five years, and is reprogrammable. (3)

At the very least both systems allow all passenger boardings to be recorded and enable even occasional users to move away from cash fare payment. The reader or sensor could be positioned to permit two-stream boarding - the driver continuing to handle cash sales - thus giving further reductions in total boarding times.

A particularly important role for magnetically encoded tickets may be provided by the requirement, within a newly deregulated UK bus industry, for a system of recording the use of concessionary passes. The use of tokens would obviously be retrograde, particularly in terms of boarding delays, and a system based on magnetically encoded cards or tickets has many attractions that go some way to justify the £40m-£50m investment required to equip the fleet. (This estimate assumes £1000 per bus - it might be somewhat less if buses already equipped with ETS require no more than an additional 'swipe' reader and connection equipment).

A further possibility raised by the use of magnetically- encoded cards is that of cards usable on more than one mode of transport - we return to this possibility in a later section.

A feature of all such systems is that the management information needs would not be changed by the fare payment system as such. Software development already undertaken should thus be largely applicable to the new systems. Whatever fare collection method is adopted, it is essential that clear aims be set for management data.

PARKING SYSTEMS

Traditional Systems

Parking charges, either as a source of revenue and/or a means of demand management, are well established. Off-streat parking fees can be collected by an attendant, a 'pay and display' system, or (with a simpler structure) an automatic barrier. On-street fees may be collected manually, but this method is likely to be labour-intensive and cumbersome, especially where much parking is short-term. The first mechanical parking meters were introduced in the USA in 1935, followed by the first in Britain (London) in 1958. Kerbside parking meters are generally simple mechanical, or electro-mechanical, coin-operated devices. 'Pay and display' equipment may be more sophisticated, enabling time of issue and varying charge rates to be incorporated. Although normally associated with off-street parking, 'Pay and display' operation may also be used on-street, eliminating the need for a meter at each bay, and hence allows a less rigid limit on the number of vehicles parked alongside a given length of kerb (where, however, one of the main objectives is to limit the total number of vehicles entering an area, an increase in spaces may not be considered desirable).

Electronic Systems

The first use of microprocessors in production parking machines in Europe was by Ticfak (now a subsidiary of Almex) in 1977. Electronic systems offer a considerable number of advantages over their mechanical counterparts. These include:

* Facilities for accepting more denominations of coin.
* Improved reliability and availability, reducing the level of servicing required and maximising potential revenue.
* Ease of tariff change (a costly exercise on mechanical machines).
* Possibility of charging differential rates by duration of stay, time of day or day of the week, including provision for free periods, and advance purchase.

Further advances in electronics have led to the testing of stored value systems for both meters and pay and display equipment, using magnetic cards pre-encoded with a fixed amount of money or time-equivalent units. Equipment is now on test using the major credit cards in pay and display machines, in some car parks. These systems help improve security by reducing the cash stored in a machine vulnerable to vandalism and theft. Costs are cut as the frequency of visits required to empty the vault is reduced. Card systems have the additional benefit of being able to differentiate between groups of users - for example, issuing discount rate cards to residents or the disabled. The range of possibilities has been outlined by Havers (1983).

Despite the potential advantages of a card system it is still felt necessary to provide for a cash alternative in most applications at least during a transitional period. This is seen as the most equitable arrangement, particularly for infrequent users or visitors to an area. Widespread use of credit cards, however, may in coming years render such concern superfluous.

A Source of Data

Most current electronic machines have a simple audit facility to provide a check on revenue collected and enhance security. It is possible to expand on these capabilities, interrogating each machine to obtain the profile of parking at a particular location including turnover, time of peak demand and average duration of stay. This information can greatly assist in estimating the effects of alternative tariffs as well as planning the optimum time and level of enforcement.

Given suitable computer software it should be possible to analyse data for large urban areas, giving traffic engineers the opportunity to practice sophisticated demand management, with localised charging structures to spread demand in terms of both time and location. Microprocessor-controlled parking machines may also be linked to a central controller for real time analysis of occupancy at each site. As a result drivers can be informed, by variable message roadsigns or route guidance systems, as to the availability of parking spaces in an area in advance of their arrival.

FUTURE DEVELOPMENTS

The flexibility that electronic machines offer is clear though as yet there has been little attempt to translate these ideas into reality, either through political inertia or legal constraints (especially in respect of on-street systems). Manufacturers have been given little incentive to develop the full potential of the new technology and have thus concentrated on improving the performance of their machines to meet existing practice.

As similar technology is developed for both bus and car parking use, the prospect of using a common card emerges. For example, the local authority operating car parks may also run its own bus system, and the same card could be sold for use on both. Some schools of thought in transport planning point to this particular advantage of thereby placing the perceived cost for parking and bus use on a similar basis. Recent, if belated, evidence of co-operation between two British Standards Institution Committees concerned with producing standard specifications for cards and readers/encoders (one dealing with parking systems, the other with magnetic card technology for bus ticketing) provides some grounds for optimism.

A further possible development lies in the experiments now being conducted into EFTPoS (Electronic Funds Transfer at Point of Sale) by banks and building societies. Here, cash or cheque transactions at retail outlets are replaced by the purchaser's personal encoded card being read, and a deduction made automatically from his account. It is questionable, however, whether such institutions would wish to see such small transactions as bus fares and parking charges on such a system (although long-distance travel would certainly be appropriate). A more realistic approach would seem to lie in using EFTPoS to purchase public transport/parking cards for use in a defined local area (credit cards such as Access can be used already to buy travelcards, etc.).

Such developments might find greatest favour in a society where credit cards are widely accepted. The UK, with some 16 million cards among a population of 55 million is behind the US in this respect but in advance of some other European countries.

CONCLUSION

New technology offers many exciting opportunities both for public transport and parking operators, and ticketing equipment manufacturers, but it is essential to have adequate software to exploit the full potential of such systems for management information purposes. Insufficient attention has so far been paid to this aspect.

ACKNOWLEDGEMENTS

We are grateful to the operators, manufacturers, local authorities and other bodies who assisted in our study, through responses to interviews, questionnaires and specific queries. The study could not have been undertaken without their help, nor that of the SERC in funding. Thanks are due also to Peter Guest of the GLC for many helpful comments on car parking systems.

REFERENCES

'Computers in Coaching'. Coaching Journal (London) May 1985, pp 27 - 29.

Havers, G.E. (1983) 'Opportunities for new technology in parking meters'. British Parking Association Seminar, Brighton, April 1983.

Mellor, A.D. (1985) 'Electronic Ticketing Systems: Ticket to improved management in the bus industry?' Polytechnic of Central London, Research Report No 11, March 1985. Available from Transport Studies Group, PCL, 35 Marylebone Road, London NW1 5LS (Price £10 UK, £15 overseas).

White, P.R. (1984) 'User response to price changes: application of the threshold concept'. Transport Reviews, Vol. 4 no. 4.

EDITORS' NOTES

(1) The multi-purpose use of local microcomputers (in contrast to a reliance on centralised data processing) is advocated in several contexts. Chapters 7, 10 and 12 mention it in the context of the rail, bus and travel industries respectively.

(2) The role of expert systems in such circumstances is described in Chapter 15.

(3) Such systems are clearly analogous to those envisaged for electronic road pricing schemes, as described in Chapter 3.

Information Technology Applications in Transport, pp. 105–140
P. Bonsall, M. Bell (Editors)

Chapter 6

INTEGRATED TRAFFIC MONITORING AND CONTROL

Job Klijnhout
Rijkswaterstaat, The Netherlands

This chapter considers road traffic control and management systems. The first part gives a general description of the possibilities to study and optimise traffic control off-line. This is followed by a description of some operational control systems.

OBJECTIVES OF TRAFFIC CONTROL SYSTEMS

Developments in electronic technology offer automated traffic control systems which are so powerful that a variety of traffic management objectives can be achieved. What in a comprehensive system such objectives should be is not just a matter of taste; in a way it is also a question of time and situation. In the old days traffic was an entity for which sufficient capacity should be created either by building roads or by capacity control measures. This attitude has been overtaken by the "selective priority" approach in which a distinction is made between different users, one often privileged group being public transport. The latest political developments show another differentiation, the need to ensure accessibility of industrial sites.

Parallel to these developments, the ways to describe the impact of traffic management evolved. To measure the benefits several yardsticks became available, the most interesting ones being:

Safety. That traffic control could be used to make traffic at intersections safer was in most cases secondary. For traffic control nowadays so much is known about the safety aspect that every country has a set of rules for traffic control installations, ensuring both safe operations and safe control. Improvement of safety as a reason for traffic control has got considerable attention in the Swedish handbook for traffic control (Signalhandboken, 1982). On high speed roads safety is a major, if not the major, objective for traffic control systems.

Delay. Average delay has played a leading role in urban traffic research. Latterly various groups like pedestrians, cars and public transport can be treated separately in the study models, so it is possible to match the study results with what the politicians need. Average delay has however a limited use. Road capacity-demand problems are similar to telephone traffic problems, where concepts like the delay value which is not exceeded 95 percent of the time, are considered. There are two classes of road user, namely pedestrian and public transport, where the 95 percentiles are of particular importance. Studies show that schoolchildren tend to ignore the signals when these let them wait for more than 1 minute. Delay values which vary a great deal make it difficult for public transport to keep to schedule.

Travel-time. This is, in networks, a good companion to delay, together with speed. The last three basic objectives are closely related, travel time and speed being alternatives, and delay forming the "waiting" part of travel time.

Density. The number of vehicles on a link or road section is experienced more directly than the flow.

Stops. The number of vehicles stopped or the number of times vehicles stop is a parameter used for three different reasons. People favour a control mode that does not force them to stop. In simple study models where vehicles either drive at full speed or wait, the stop can be used to account for acceleration delays. More recently, stops have been used successfully as a basis to calculate the pollution and noise generated.

Queue length. This can be used to forecast the risk that queues will block an intersection or access, but it is difficult to measure, both in practice and in simulation models.

Fuel consumption. During the energy crisis plans for reducing fuel consumption suddenly became very popular. Although a parameter used for evaluation, fuel consumption is normally calculated on the basis of travel time and stops.

Environmental aspects. Examples used are noise and pollution. Both are, like fuel consumption, derived from speed, delay and stops.

How far one or more objectives can be achieved is a matter of knowledge and equipment. Three strands or philosophies in the research into traffic control and signalling can be distinguished.

One is the straightforward approach. A traffic control program is designed to achieve a specific type of control strategy. The idea behind it is that the strategy specified will, by picking out the most important aspect of traffic control, lead to an efficient control mode. The best example is the so-called green wave, where traffic lights along a corridor or arterial are linked in such a way that traffic can, once having entered the corridor, travel along it without delay at the intersections.

Another strand is the analytical one. By studying the effect of different control programs in various situations it should be possible to learn how to define the best control program for a given situation. A further extension leads to a general set of rules about how to calculate a control program. Programs will be selected on the basis of knowledge about their effects.

The third idea is to let a computer run an on-line optimisation program to define what, at each moment, the signs and signals should show. The optimisation program uses a simulation of traffic and control to forecast the effect of possible traffic control measures.

Developments have given us various tools for research in the above three areas. Some of these will be discussed in the following paragraphs.

INTERSECTION CONTROL

Study tools

Although there were attempts in the past to find analytical solutions to the optimisation problem of local intersection control, most of the research is based on the results of simulation models.

It is interesting to see how gradually these models became more refined in simulating traffic behaviour. For an accurate representation or simulation of the effect of a signal control plan in which green phases can be extended on the basis of detector information, the simulation of the trajectories of vehicles should be detailed enough to reflect deceleration and acceleration. Also vehicle behaviour should be described in time steps of 0.1 sec. rather than 1 sec., the value which was used in early models. In studies we found that these refinements were necessary to simulate correctly the difference between control with standard size loop (1) detectors and long loop detectors (with a loop 20-40 m long).

The effect of the use of queue length as an input in a closed loop operational control mode is very difficult to simulate as it is difficult to simulate traffic sufficiently accurately in queueing conditions. Writing a simulation program to simulate isolated intersections is simple, writing a good one has turned out to be quite a task. It is therefore not surprising that many simulation models have been published, and that at the same time there is hardly one which is thoroughly validated against a sufficiently reliable set of data collected at an operation site. An example of a good model is SIGSIM (1980), made for the Swedish Ministry of Transport, which is currently believed to be the only validated model.

When one wants to use vehicle actuation in a special control mode for public transport vehicles, the models still let us down. The model FLEXSYT-II by Middleham (1983), to be available soon, should offer all necessary research facilities for such studies. Looking at the detection and classification possibilities, the development of study tools is out of phase with the hardware.

Simulation models were in most cases used to study a specific situation. However, in some extensive research programs, the results were used to find general rules for signal plan timings. A famous example is the cycle time and split rules of Webster. (2) Simulation studies have also led to some general conclusions. As an example, it was found that the higher the traffic demand on an intersection, the more rigid control should become. Maximum green times are then more important than extension timings.

Such conclusions have led to the design of BASICSPEC (Basicspecificatie, 1980), a structured traffic control program for vehicle actuated controllers. BASICSPEC plans are made to ensure that under both light and heavy flow conditions the right combination of traffic flows will get green. Thus, on one hand the highest flows will be activated in the right sequence without the danger that minor flows can disrupt this sequence, on the other hand interleaving of flows is as flexible as possible avoiding unnecessary delays common to any rigid sequencing. BASICSPEC is used by the Dutch Ministry of Transport.

Operational systems and on-line optimisation

In the past 10 or 20 years manufacturers have made equipment sufficiently flexible to cater for all kinds of control philosophy needs. This has led in different countries to different developments. Most countries have restricted themselves to a simple form of vehicle actuated control. In the Netherlands, however, with its tradition of bicycle traffic and its political decision to promote public transport at the expense of other traffic, traffic engineers have tried to make full use of this flexibility. Basically the Dutch controllers are signal group controllers; a signal group being a set of traffic signals which at all times show the same aspect (colour or colour combinations). The control program defines when each signal group should change. All control decisions are written in terms of signal group status and detector status. In these controllers not only are most traffic streams treated separately (have their own signal groups), but also cyclists, pedestrians and public transport have their own signal groups. This makes the control equipment fractionally more expensive, but also much more flexible. There was a disadvantage to this freedom; it has taken long to learn how to use it properly. BASICSPEC is a good example of how to manage flexibility.

The developments in Sweden should also be mentioned where already at a very early stage a similar refinement, namely the splitting up into separate signal groups, became common. The Dutch BASICSPEC scheme to manage diversity is very similar to control strategies used in Sweden.

All in all we may say that precalculated vehicle actuated control schemes are standard for isolated intersections. But we can already see a new era of self-optimising controllers.

More than 20 years ago Miller (1965) developed his model for a two-stage controller which calculated at regular short intervals which of the traffic streams should keep or get green. The calculation is based on actual traffic data measured by detectors, and is designed to minimise overall delay. The results of this theory were successfully tested out by Baang (1976) in his TOL strategy. At the TRRL, Vincent (1986) has made the model operational by creating a solution to the instability problem. It has been christened MOVA.

An American development, the OPAC scheme of Gartner (1982), uses dynamic programming techniques to calculate the time settings. The results so far put this program in the lead.

Oversaturation

Oversaturation manifests itself by the (ever) growing queue of vehicles which cannot be serviced. Where such queues are a real nuisance, the old control objectives have to be replaced by new ones like:

- use the intersection capacity as well as possible, leading to what is known in control theory as bang-bang control. Bang-bang implies using extreme control parameter values only. Here the signal groups handling most traffic per second get maximum green at the expense of all others.
- manage the queues such that the ratios of the queues on the arms of the intersection are kept constant.

Although the problem seems to be one attractive to mathematicians, the theoretical developments so far have contributed to a better understanding of the problem, but not significantly to solutions. A weak point is still the identification of oversaturation in time. Queue detection mechanisms can work reliably, but then they will only react when oversaturation is definitely there. A preventative approach is preferred to a curative one, let alone a curative one which is activated too late. The OECD report of 1981 gives a good overview of the problem.

URBAN NETWORK CONTROL

Study tools

To co-ordinate or not to co-ordinate, that is the network control question! This can only be answered when one knows the effects of co-ordination in terms of improved performance, expressed with the help of the objectives discussed above and knowledge of equipment costs.

Looking at the way network control is set up we find the same three schools of thought as found for the isolated intersection:

- the simple solution like the green wave, supported by many;
- the use of optimisation models like TRANSYT to calculate the (3) offsets (i.e. the co-ordination);
- the self-optimising systems working on-line.

Green wave optimisation

Mathematical models to calculate the maximum bandwidth of a green wave in two directions have been known for some time. Their handicap is that a diligent traffic engineer can easily obtain better results by accepting slight changes in the green wave speed. An attractive development in this field is the MAXBAND model, creating a maximum bandwidth along an artery, accepting a change in green wave speed and taking care of flow changes at the various intersections.

Off-line design of proper green waves is thus possible. Whether these green waves give the solution wanted is not only a matter of taste. There is some evidence that the green wave (apart from its advantages, like being a control well accepted and understood by both politicians and drivers) is a mode which all in all scores high when applying the objectives mentioned above. The results of comparative tests carried out with the help of NETSIM, an American network simulation program, are interesting. Delay figures with settings based on the MAXBAND principle could match those yielded by TRANSYT's optimisation program, because it is the major arterial which determines the signal settings.

Off-line study tools

The traffic network study tool TRANSYT is the best seller among traffic research programs. It simulates traffic in a network with signalised intersections, calculates the delays encountered by the simulated vehicles and uses a hillclimbing optimisation program to adapt the signal timings (split of green time between stages and offset between adjacent intersections) such that overall delay is minimised, all for a given traffic demand. Optimal control schemes found this way, either using TRANSYT or its American counterpart NETSIM, are tailor-made for the situation studied.

This leaves problems of two different sorts to solve:

- What makes a traffic situation typical (that is, how can we detect and define the actual traffic situation), how many different situations should we cater for and how do we move from one control scheme to a more appropriate one when changes in traffic require it?
- Can we deduce from the control schemes TRANSYT calculates a direct relationship between the data describing the traffic demand and the required signal settings?

Concerning the first point, the most used solutions are:

- Rely on the capacity of the network to cope with any normal traffic demand and try to reduce delays by local vehicle actuation within ample maximum green times. This ignores all that could be achieved by proper co-ordination, but it is cheap.
- Calculate control plans for various traffic situations and find reasonably representative traffic parameters which can be used to recognise these situations, selecting then the plans on the basis of actual traffic data. A potential problem is that traffic flows can increase so rapidly that the mechanism to detect flow changes does not react before it is too late. In a following section, a computer model to analyse the motorway traffic situation is discussed. A similar approach can be used to analyse the urban traffic situation and to decide when to change pre-programmed plans. As far as plan changing is concerned, a straightforward all over change is preferred. In those cases where plans are selected which offer some spare capacity rather than a sharp minimisation of delays, the above problems can be avoided.

Concerning the second point, the relationship between the traffic situation and the network control program, some very general rules for urban traffic control have been found. The split of a cycle at an intersection can be calculated on the basis of intersection data only. The denser the network and the higher the traffic demand the more the need for co-ordination. Under these conditions, vehicle actuation becomes more and more constrained by the rigidity of the co-ordination. This also gives an indication how to decide when to leave control to the local controllers and when to enforce co-ordination.

Application of these rules is possible with various off-the-shelf systems. The level of detail in which the manufacturer has already laid down the decisions differs, however, very much between the firms.

Simulation models, like TRANSYT, make it possible to study the effect of co-ordination in networks, but they are not suited for a study of network control with vehicle-actuated controllers or public transport priority. With some clever fiddling, calculating per intersection what spare capacity is needed to suit the public transport priority treatment, and then optimising the offset one can find a solution for this problem. It would, however, be far better to have a proper simulation model suited to this problem. There are plans to introduce such a model, called FLEXSYT-II, in 1986.

On-line optimisation

Finally the third philosophy: do all the work on-line and let a computer program more or less continuously decide what level and kind of co-ordination is needed. This brings us to the network alternative of TOL, MOVA and OPAC, namely schemes like SCOOT, SCAT and UTSC, about which more later.

Oversaturation

The major problem with oversaturation is that if one or more sections of a network get congested, the effect is often cumulative and very rapidly the whole network gets jammed, reducing the throughput of otherwise unaffected traffic to practically zero.

Oversaturation merits specially adapted study tools since oversaturation control programs differ greatly from those applied under normal circumstances.

Analytical models have been developed to find the control

strategies to handle temporarily overloaded links. Various studies, such as the one in Bordeaux, indicate that a very strict metering of the traffic entering the network should be applied to avoid an overload of the bottleneck intersection or area. Specially adapted off-line simulation models like THESEE-SATURATION (Texier, 1980) can be used to study the effect of metering of traffic at different entry points. Simulation models originally meant to study the effect of traffic measures on the redistribution of traffic flows through a network, namely CONTRAM (Leonard et al, 1978) or SATURN (Hall et al, 1980), can also be used. However, Origin-Destination information is needed to get good insight into the various traffic streams involved. It is costly and time consuming to get correct O-D data.

From various test cases it has become clear that, as for the isolated intersections, actions should be preventative. However, as with the plan selection problem for normal traffic, finding the right traffic parameters to detect the onset of saturation is difficult.

A special problem in oversaturated networks is the effect on public transport. Often people try to reduce the delays for public transport by special measures at the intersections or on some links. This can make matters much worse for the rest of the traffic, creating a vicious circle. First of all, the network should be kept running; then in that network, but best of all at its entry points, one can give public transport preferential treatment.

Operational systems: plan selection

From the various schemes to control networks which are in use one can conclude that the traffic engineer in general lacks the manpower and money to set up and run a sophisticated urban network control scheme. The majority of schemes operate a plan selection system working with a modest number of detectors. Such schemes have been in existence for more than 20 years. The detector information used was mainly the number of vehicles on links to (or in) the network.

To cater for the difference in vehicle length, occupancy values (percentage of time a detector has a vehicle over it) are used. Decision trees working this way have long been standard with, for instance, Siemens. Very interesting results were obtained in Stockholm where tests showed that speed data should also be used to classify the traffic situations. Where to measure what still seems a matter of taste. From experience, one can conclude that data both on the efficiency of the applied (local) intersection control as well as on the actual performance of the major links are needed.

The architecture of urban network control equipment has changed considerably in the past 20 years. The first series used a very centralised sort of control, the local controller equipment being nothing but a lampswitching slave. Gradually the local controllers got more room to manoeuvre and only overall information like plan selection and co-ordination remained central. This change has two advantages. First, it requires less communication between the centre and each controller. Second, it makes the system less vulnerable to communications failure, since a reasonable control is ensured by the intelligent local controller. This is a very important aspect in any control system. Communications should enable an adequate exchange of information, but the distribution of intelligence should also ensure that the system as a whole cannot easily fail catastrophically. This aspect is a design concept labelled graceful degradation.

Plan selection has also changed. In the old days, simple decision trees were difficult enough for the traffic engineer to program. Now decisions are made in a more stepwise way. Information from one or more controllers about traffic as well as about the operational control plan is used to select what co-ordination and control plan to use next. Such decisions can then be combined areawise for overall decisions. These selection procedures can be offered now by many manufacturers with such a level of detail that it becomes very difficult to program them for operational use.

It is one of the most misleading aspects of these schemes that although they offer the traffic engineer a very flexible selection mechanism, they at the same time impose a certain decision mechanism, the consequences of which he might not realise. As there is still an enormous gap between the off-line study tools and the dataset to program such operational selection schemes, it is very difficult to find out how, and on the basis of which information, control plans should be selected, and further, how to translate such a selection procedure into a decision program offered by the manufacturer. Facing such fundamental and practical problems it is no wonder that many simply skip these questions and program their system as they see fit, with an acceptable though probably not perfect control as the result.

Finally, there is also a very practical reason to go for central control, namely the facilities central control can offer for servicing of the equipment. This will lead to its own demands on the communication between the local equipment and the central computer. Experience has shown that such equipment monitoring facilities when integrated in a network system do pay off.

Concluding this section, we can say that central plan selection and co-ordination is the most common type of network control with the following advantages:

- There is factory-made software to decide which plans and which co-ordination to apply.
- Intelligent local controllers are used, which, especially under low flow conditions, can run flexible, vehicle adaptive programs. When traffic demand increases more (co-ordination) restrictions will be imposed centrally, which will tighten control.
- It offers data for servicing management.

One example of such a system will be discussed in the next section.

The SCAT system

SCAT, which stands for Sydney's Co-ordinated Adaptive Traffic system comes from the Department of Main Roads, New South Wales, Australia. (SCAT system, 1979). The hardware comprises microprocessor local controllers, mini computer regional controllers for up to 120 local controllers and one central monitoring computer. The software structure of SCAT in the regional controllers works with local controllers which are grouped in kernels or sub-systems. These sub-systems in their turn are grouped in systems. Kernels belong to one regional computer only.

Monitoring of the whole network is possible in the central computer. Smaller cities might do without it and use not more than one regional mini computer. Traffic co-ordination and actual control are not so centralised. It is implemented in three levels. The lowest level is that of the intelligent local controller. The next level is that of the sub-system or kernel. Decisions on this level are made in the regional computer. The highest level is that of the system. Note that this computer can manage more than one system (of sub-systems) and that there is no control interaction between these systems. All in all, this set up is a logical result of the control philosophy and of the equipment maintenance requirements.

The control philosophy is a bottom-up one. On the basis of traffic data, plans for the local controllers and the offsets between the controllers within one kernel are selected (if traffic demands require such a coordination). High traffic demands can lead to the selection of plans which link the sub-systems of one system together. All in all, SCAT is essentially not different from what for instance the standard Swedish equipment can do. I see it as primus inter pares because so much is published about the results. SCAT is outstanding because of the way quite elaborate, though straightforward, procedures operate centrally to decide which cycle time all controllers of a subsystem should use, which constraints on the splits each controller should apply and what co-ordination (offsets) should run.

The SCAT System uses degree of saturation, which gives an idea of the efficiency of the use of a green phase. In general, DS is the ratio between the traffic demand for green and the amount of green available. The effect of vehicle length is taken into account. DS is measured on all strategic links. Note that it is data about traffic passing the intersection's stopline, not direct information about the traffic to be handled or about the situation on the links itself.

In principle, full vehicle actuation is used by the local controllers. Phases may be skipped or extended as traffic data demand, all within the sequencing restrictions of the operational signal control plan and within the restrictions imposed by the regional computer.

One can see from the mechanism that in the case of oversaturation the system will run more or less fixed time control which in case of an adequate plan should be capable of handling all traffic that is in the network. When properly designed, this will thus mean a metering of incoming traffic. There is, however, one extra feature to avoid spillback effects inside the network, namely the use of the system level selection mechanism to detect queues on spillback prone links in time to react and select a special oversaturation system plan.

Summarising, one can see that SCAT offers a neat instrument to gradually enforce coordination restrictions when needed, that a lot of experience is reflected in the choice and use of the DS parameter, but that there is still room for improvement on the level of kernels. Special simulating facilities are therefore being built in Australia to get a better link between off-line study and on-line application.

Operational systems: on-line optimisation

In urban network control we differentiated between green wave systems, plan selection systems and on-line optimisation systems. SCOOT (Hunt et al, 1981), discussed in this section, belongs to the latter group, meaning that it has its own mathematical model to represent the traffic situation plus its own signal setting optimisation program.

SCOOT is essentially software which, on the basis of actual traffic data, calculates the required signal setting. The idea is that an already existing urban traffic network control system can be upgraded by adding SCOOT as a brainy black box. However, some hardware changes will also be necessary because urban network control systems do not normally collect the quantity and type of traffic data SCOOT requires. The philosophy behind the SCOOT model is that the model should regularly simulate the effect of various traffic settings and then decide which traffic setting should be used. A very simple description of it could thus be an on-line TRANSYT. There is, however, more to it than just running a fast version of TRANSYT every, say, 4 seconds. How can SCOOT be designed to react in time adequately to changes and how can drastic plan changes be avoided?

Little is published about the workings of SCOOT. It is owned by the British Department of Industry and only a (happy) few British firms and the Godfathers at TRRL know the source code. From the test publications one may conclude that, in many circumstances, SCOOT functions very well so we will try to give some more information in the following paragraphs.

SCOOT tries to forecast the effect of traffic signal settings. In order to do this, it collects data which will give the longest forecast span possible. Therefore, instead of measuring traffic which crosses the stopline, reacting on the traffic light settings previously calculated, it measures traffic which is approaching the stopline. Detectors are placed as far upstream as possible from the stopline, directly below the previous intersection. This is done on every link, using the TRANSYT term link for a route between two junctions or nodes. These link detectors are scanned 4 times per second. Normally 1 loop detector is used per 1 or 2 lanes.

Traffic crossing a stopline is normally not measured. This means that there is no easy way to cater for temporary changes in the saturation flow. Only indirect ways are used. They are a check on the link detector for the next intersection, which acts as a junction exit detector, and the use of a special detector upstream of the stopline to test whether the queue has reached this point or not. It is interesting to note that in SCAT so much emphasis is put on a correct measurement of the saturation flow.

All detector traffic counts are collected over short periods (the literature mostly speaks of a few seconds, let us assume here 4 sec. periods). They are used to forecast when traffic will arrive at the stopline, either to cross or to queue. For longer links the effect of platoon dispersion is taken into account. Thus measurements could give quite accurate short term predictions or information about traffic at a stopline. SCOOT, however, has more than this short term information; it also stores information collected in the past. This is done for a period or window of just over 2.5 minutes, say 40 steps of 4 seconds each. Thus SCOOT has for every stopline a 2.5 minute overview of arrival patterns, of which the very first part is a forecast based on the latest link detector information (inclusive of a platooning effect), and of which the rest is an expected arrival pattern based on historic data and used as forecast. As time advances, the information from the link detector is used to adapt in a smooth way the stored arrival pattern information, the expected arrival pattern is thus an exponentially smoothed average. This combination of the average arrival pattern and the more random actual flow values is essentially all the traffic data SCOOT requires.

Comparing SCOOT in more detail with TRANSYT, one can say that the ideas about the network model, with its links and its nodes, as well as about traffic behaviour in terms of arrival patterns and platoon dispersion are the same. The optimisation, however, is completely different.

SCOOT runs three different optimisation programs at different intervals. Thus decisions are made at three different levels (in the same machine) only when necessary.

Splits may be changed by a few seconds at a time. Therefore, only just before a traffic light change is due a check is made on the basis of the degree of saturation of every link of a node whether the split should be changed or not.

Once per cycle, during a given phase, a check is made whether the offset should be changed. Only minor changes are allowed, so the check can be quite simple; which way to change and by how much. Some off-line models do the same. Offset changes will have an effect around the node itself. This is taken into account because SCOOT works with arrival patterns which, when such calculations are made, will be adjusted.

SCOOT works with pre-set combinations of intersections. These sub-systems are independent. Within a sub-system, all nodes have the same cycle time or half that value. The decision to run full cycle or half cycle is done when the cycle length is checked, with a restriction against frequent changing. This is done at time intervals equal to or larger than the arrival pattern information block which is some 2.5 minutes long. Within each sub-system, SCOOT looks for the most loaded node and calculates which cycle length would give a 90 percent degree of saturation. This is quite a high value, leaving little breathing space, which shows how confident the users dare to be in SCOOT. Cycle lengths are adjusted when needed, but with the restriction that changes are limited to a few seconds only.

SCOOT has little provisions to handle oversaturation. The way the queue information from the link detectors is used has disadvantages. SCOOT will, once a queue is detected, try to get rid of it by giving more green at the downstream stop line. In more general terms, SCOOT will tend to minimise delay, which is not necessarily the same as ensuring a maximum throughput. The time span SCOOT uses for prediction is too short to calculate the effect of metering. So like most systems under overloaded conditions, the maximum values for various timings given beforehand will define whether or not the network can be kept running and in the case of SCOOT some special action should be taken to ensure that especially the split optimisation works with criteria more suitable for saturation.

The difference in approach between green wave plan selection and on-line optimisation is great. The difference in results is not. A lot of expertise has already been built-in in many of the available programs, but still significant expertise and effort is needed to get the systems running properly. What, however, makes on-line optimisation stand out is that once installed the system stays up-to-date which is not the case with the other models. Little is known about the losses caused by not updating fully vehicle actuated local controllers or fixed time network plans, but they are certainly a few percent per year. (4)

MOTORWAY SIGNALLING SYSTEMS

Developments in electronic technology offer various automated motorway signalling systems for a variety of traffic management objectives. The majority of these systems are dedicated to one (5) task, being in most cases either ramp metering, fog warning or queue protection.

The objectives for motorway control are in general terms:

- more efficient use of the available road capacity;
- improvement of road safety;
- easing of the task of maintenance crews and of the police;
- collecting research data.

Compared with the urban traffic control objectives not only the third on the list is new, also the others differ when we go into more detail.

When we take a closer look at how one can achieve these objectives, we can define a set of tasks, for which in turn we can specify the necessary data, data processing and actions. There is a variety of overlaps between the various tasks. This, combined with the cost figures, makes it possible to define which combination of tasks can be carried out cost effectively by a system in a given situation.

Safety

Secondary accidents. A survey in the Netherlands showed that on the motorways with a traffic load of over 60,000 veh/day about 50% of the accidents are caused by severe disruptions in the traffic flow, caused themselves by accidents or queues. Furthermore, the risk that once traffic is disrupted a secondary accident will take place reduces practically quadratic with time to become virtually 0 after 20 minutes. This leads to the conclusion that serious disruptions should be detected very rapidly. Practical experiments have shown that once a police car on the hard shoulder is flagging traffic down speeds are reduced to 70 km/h and traffic will come to a standstill in a smoother manner. With such queue protection there are practically no secondary accidents. This was behind the fully automatic queue protection system operational since before 1980 in the Netherlands, which has reduced accidents by 25 percent. The system advises drivers approaching a queue first to slow down to 70 km/h and next to drive at 50 km/h. The 50 is shown very close to the tail of the queue as well as in the queue. Showing a 50 over the queue proved to reduce the speeds in the stop and go mode by some 20 km/h, and increased the throughput by , in these circumstances, a very valuable 4-5 percent.

Safety in adverse weather. Wet or snowy roads are 3 times more dangerous than dry roads. Detection of adverse weather situations is not simple. The tendency is to try to predict ice and black ice in time so that the road can be gritted, sanded or sprayed with salt. To do this, quite an extensive measuring program is needed, as was shown in international co-operation by EUCO-COST-30. Installation of a system can not only have a positive effect on safety, it will reduce the environmental consequences of winter maintenance.

Sites where black ice is often the cause of accidents can be equipped with simpler special black ice detectors which switch on alert signs upstream . In such cases it seems necessary to tell the driver that there is a risk of black ice by using the appropriate international symbol; the ice crystal. With black ice, advisory maximum speeds often have to be so low that without an explanation they could well be ignored.

Safety in fog. Fog accidents tend to make newspaper headlines. The pile-ups have often been spectacular and horrifying. Still very little has been done to avoid such accidents successfully. Experience with the Dutch Motorway Control and Signalling System (MCSS), although it will take some more years to satisfy all statisticians, have been sufficiently encouraging to mention here.

Detection of fog is possible, detectors which measure how much light is scattered by the fog give the best results with the least practical problems. However, dual carriageway roads require detectors every kilometre at both sides for a faultless coverage, which is not feasible on cost grounds.

Traffic speeds in fog are generally too high to avoid an unexpected lane blockage. Speed restrictions in fog will, to give no legal problems, have to be low enough to avoid such a blockage. Such a speed turns out to be unacceptable for a fair proportion of drivers, which leads to different groups of drivers driving at different speeds in fog, an unsafe situation. In the Netherlands for a period of a year, advisory speeds of 60, 70 or 80 km/h were shown in fog, the one equal to or just below the actual average speed being selected. This led to a reduction of the variation or distribution of speeds. However, the speed shown was felt to be too high, and most importantly, the difference between speeds of such a blanket measure and speeds set up to protect a queue (70, 50) was not obvious enough to alert drivers of queues. Analyses of fog accidents in the meantime led to the conclusion that typically they start as traffic jams without any (severe) collisions, followed by a series of head-tail collisions of which one or more can be very severe in terms of fatalities and casualties. Since in MCSS queues are detected and protected automatically, the decision was made to do nothing special in fog and leave it all to the Automatic Incident Detection (AID) mechanism. This has up till now worked to our entire satisfaction.

Figure 1 shows the number of queue protection actions for a typical morning fog situation involving some 200 gantries (or in American: sign bridges). Data are given for ten minute periods. Curve W shows how often per period MCSS decided that at a site traffic conditions changed between queue and free flow. The number of 50-70 queue protection measures varies, maximum and minimum values are given by curves M and L respectively. It is clear that in such circumstances human control of the signalling measures would be impossible.

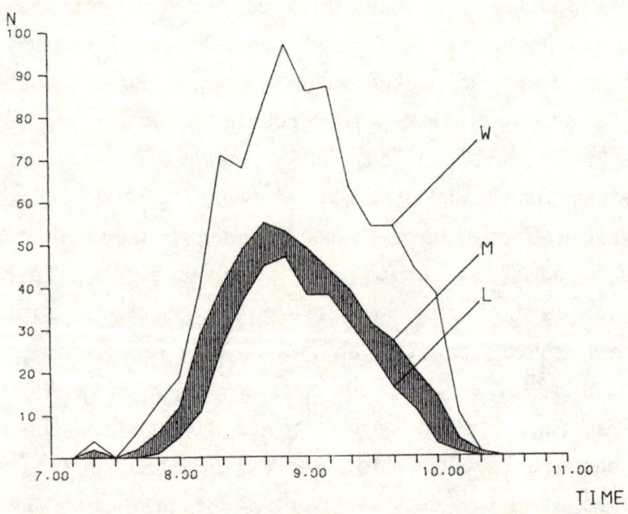

Fig.1 Number of AID measures in fog

Delays. Delays on motorways can be especially severe in the case of recurrent overload and in case of roadworks and accidents. To reduce these delays different schemes are available.

Recurrent overload. In this case, the standard management solution is to keep traffic running by keeping the flow volumes below a certain breakdown level. When traffic flows exceed that level, shockwaves will sooner or later bring traffic to a halt. Once stopped, throughput figures will never become more than about 90 percent of that breakdown value until traffic can flow freely again. To detect where this breakdown value lies, two algorithms can be used. One measures at those points where in most cases shockwaves start, the other measures traffic entering or leaving a motorway section. The first algorithm should detect the shockwave prone situation directly whereas the other tries to predict traffic conditions on the motorway section. It is somewhat situation dependent which algorithm should be preferred. The first one is suited to typical bottleneck problems, the second is less site sensitive.

The action to take when there is a risk of overload is either to ramp meter, to filter traffic, or to reroute it. Especially in the USA, ramp metering is used to avoid overload. Access is restricted by a metering program which decides how long per access or ramp the gap between successive vehicles entering the motorway will be. When there are no specific bottlenecks, careful tuning of the detection and metering can be difficult. Another problem is that American experience has shown that such ramp metering is difficult to sell to the public.

In some American and Japanese situations, it is possible to use parallel roads to drive from one ramp metering site to another further downstream. This offers not only a redistribution in time but also in space.

Rerouteing is a redistribution in space. To be accepted by the drivers, rerouteing should give substantial benefits (of the order of 10 to 20 minutes), and certainly no driver should be worse off.

Finally, the whole network for rerouteing should be of the same (motorway) quality. In Germany, successful rerouteing schemes for motorway networks are operational. The rerouteing advice is given by changeable direction signs, and, unlike ramp metering, is advisory.

Smoothing. After not so successful tests to increase capacity by imposing a maximum speed of 70 km/h when traffic volumes exceeded a certain value, a more flexible scheme was tested in the Netherlands. With the MCSS, traffic volumes and speed were constantly monitored. Just before volumes reached the breakdown level at which shockwaves would start to affect traffic, advisory speeds were lit, and a speed which was just below or equal to the actual average speed was shown. The effect was that traffic volumes could increase by up to 5 percent before the shockwaves would reduce capacity. We found that there was no simple way to decide when to switch these signs on and off, there are certainly no fixed figures and weather can affect the levels by some 10 percent. After the pilot tests this scheme to reduce overload delays was set aside and it was back to the drawing board to find a good algorithm.

Road works and accidents. With the increase in traffic flow and the ageing of the motorway network, more and more small repairs have to be made. These repair works ask for a safe protection of the maintenance crew and at the same time should not hinder traffic more than necessary. Setting up and removing all the necessary signing costs on average as much time as the actual repair work. And with heavy traffic loads, a simple signing arrangement does not give a safe and smooth merging. This has in the past led to the introduction of signs of the red-cross/green-arrow type to reserve lanes for maintenance.

The Dutch MCSS is the first motorway system to include flexible lane reservations as standard. In this system a driver approaching a lane reservation will first see a gantry with a white slanting arrow over the lane which will be blocked further downstream, accompanied by speed limits (90) over the other (open) lanes; on the next gantry he will see a red cross over the lane blocked and speed limits over the open lanes. The latter legend combination is repeated over the whole length of the lane reservation. The effect is in many ways positive.

Total repair time is now reduced by 40 to 50 percent during daytime and 20 to 30 percent at night, when battery illuminated additional lane markers have to be placed on site.

Merging capacity is increased from 1400 veh/h to 1700 veh/h. If capacity is not sufficient, there will still be queues, but as these are handled by the AID mechanism throughput along the lane reservation is now 4-5 percent better than before.

There is a cumulative effect as a result of these MCSS facilities; shorter disruptions give less queues which are handled faster and in total much less traffic is affected. The reductions in delay due to MCSS give an economic benefit which pays for more than half of the MCSS costs.

Road Maintenance and Accident Handling

The third objective, to ease the task of maintenance and of the police, has already been touched on in the previous paragraph. There are surveillance facilities in many motorway signalling systems. But nearly all have a detector network which is too weak to rely on fully, and some systems only rely on TV coverage. This can be of some help, provided there is no fog etc, to find out what the problems are but surveillance of a whole network by TV is rarely possible.

Another important police task is accident handling. Although in some motorway signalling systems police can switch signs, there is evidence that such actions are only partially effective. The ergonomic requirements are not met by most of these systems. These requirements are, amongst other things, that the information should not be given more than some minutes driving away from the accident, and that the information should be correct. This requires a very good monitoring of traffic and signing. These requirements hold for any signing, but they are difficult to fulfil, especially for accident handling. Clearly monitoring traffic and alerting in the case of incidents should be done fully automatically.

There remains the actual accident handling by the police. They prefer to work with lane reservations as protection at an accident site. They also want, in some circumstances, to reserve the fast lane to clear the way for emergency vehicles. These requirements are all met by most tunnel signalling systems and by the MCSS.

Lane reservations for planned maintenance work in the MCSS have been mentioned before. Additional to the gantry sign symbols, the maintenance crews place a vehicle just under the first red cross sign to protect them against reckless drunkards (who may have made the roads, but are very undesirable as users). They also use cones alongside the reserved lane to avoid people swerving into that lane after having passed one crew working; there might be a second crew. However, for immediate repairs and for accident handling there is no time for these extras, and lane reservation is enacted by the MCSS signs only. The police reckon that for accident handling in MCSS they can reduce their manpower by half.

Research Data Collection

Motorway traffic research has been severely hampered by the lack of proper research data. Even now, most signalling systems collect only that little bit of information these systems need for their own operational use. Research requires more, more accurate and non-filtered data. In most countries, research data collection is not taken care of by a motorway signalling system but by mobile data collection equipment. MCSS is one of the very few systems which offers research facilities. Detector data from up to 16 sites can be collected and dumped on tape according to a European standard.

Study tools

Off-line study facilities are, certainly when compared with urban traffic, scarce and piecemeal. Two of these tools will be mentioned here; AAS (Hilgers 1980) to study incident detection and FREQ (Jovanis 1978) to study ramp metering. Apart from these two, there is a series of programs to simulate traffic on a motorway, but these models are useful only to those who have the facilities to tune them to make them fit the local situation.

AAS. In the research into AID we can distinguish three different approaches. First the empirical trial and error way, where parameters, which should be related directly to traffic jamming, are used to monitor traffic. The exponentially smoothed speed is an example, the relation between its value and the definition of traffic jams being quite obvious. Second the traffic theory approach, where disruptions are detected by comparing traffic behaviour measured at two (up and downstream) sites. An amalgamation of these two are the models which estimate the traffic density on the basis of up and downstream detector information and then classify the traffic situation with respect to this parameter. This fits well with the research for rerouteing. Third the factor analysis and clustering theory approach. In this case detector data is processed to see if data collected over an incident period can be distinguished from the data collected at normal flow periods by applying clustering theories.

In all but the first approach, there is an important distinction between AID algorithms made to detect just traffic jams and those meant to detect any disruption of flow dangerous enough to call it an incident. For the first traffic jam type a variety of algorithms will function alright. The effect of any incident is less under low flow conditions. A spacing of 150 metres seems adequate to detect any effect on traffic of any stalled vehicle. A spacing of 500 metres seems adequate for any traffic flows heavy enough to let an incident be the cause of a potential problem.

The effectiveness of Automatic Incident Detection algorithms can be studied with the Dutch model AAS, which stands for AID Analysis

System. This can also help to define the most effective settings of an algorithm's parameters. It is built around a model to analyse traffic data with the help of clustering theories. As input it requires the "raw" detector data as collected from the detectors along a motorway with as little pre-processing as possible because any filtering might affect the results of AAS runs. If no field data are available the detector data can be produced by the simulation model of the AAS suite, which microscopically simulates traffic on a motorway. This model, based on a model from the University of California, requires as input a description of the detector layout, of the flows and information as to when one or more cars should create an incident by stopping.

The detector data are, in AAS, used to calculate the values of various user defined parameters. These are then used to calculate the values of the features, defined by the user describing the traffic situation. When there are n features the situation at a given point in time is thus described as a point in a n-dimensional space. AAS stores all these points. The next step is to decide which points together form one cloud or "cluster". If the situation description is based on useful features, these clusters can then be labelled with a verbal description of the traffic situation. If data collected during an incident is scattered randomly over all clusters, it will never be possible to use these features in AID to detect an incident. If not, the next step will be to find simple separators (planes between these clouds) and to use these separators to section off the incident clouds of the n-dimensional space. It is possible to use AAS to transform the n features into a smaller set of m features and to rerun the rest of the process. AAS can calculate how much less effective the simpler version is in terms of wrongly classified points. The result of this off-line research, namely the definition of separators between the classes in this m or n dimensional space spanned by the features, is all that is needed in an operational incident detection model. In an operational AID program, features are calculated just as in AAS, and whenever a new value is found a simple check is made against the separators found by AAS to determine which class or traffic situation currently exists. The AAS classification and partitioning has to be simple, so linear separators are used, as otherwise the operational AID will be too time-consuming.

AAS has been used and is still used extensively to study the value of various operational AID algorithms. Also much of the knowledge about the effect of detector layout, etc., comes from AAS studies. In Figure 2 a typical example of the AAS calculation of the effect on the reaction time of an AID algorithm of, firstly, detector spacing up and downstream of an incident, and secondly, the flow values, is shown. The incident in this case is a vehicle stalled on the slow lane of a two-lane carriageway.

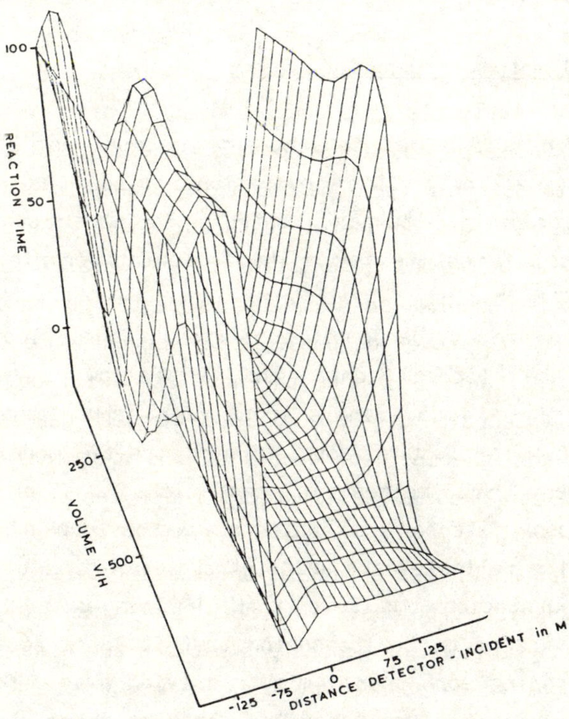

Fig.2 Reaction time of an AID-program

FREQ. The FREQ model of the University of California, Berkeley, allow for an optimisation of the use of the highway capacity by appropriate metering schemes. The motorway can be split up in sections for which the user will have to define the capacity. The program uses a macroscopic representation of traffic, and flows entering the system can be defined by successive time slice, allowing a study over a long period including possible "after effects" of metering. The program looks for the optimal ramp metering scheme with objectives as requested. Not only delay but also noise, pollution, etc., can be taken into account. The effect of various car-pooling schemes can also be studied. In fact, FREQ forms a suite of programs each suited to a given aspect. Also a link has been made with TRANSYT and the effect of ramp metering on the underlying network can also be studied.

Operational motorway control systems

The Dutch MCSS is at this moment the most complete system for motorway signalling. It is operational since 1982 and a Mark-II version comprising tunnel control, tidal flow and general improvements, is on the way. The following description is of the original MCSS version.

Detectors are of the double loop type. Double loops are used to get accurate (speed) data and to allow a check against malfunctioning. Also, single loops might shut the door for the implementation of future, sophisticated traffic models. Loops are placed every 500 metres in each lane. Each site has a microprocessor detector unit which scans the loops, checks the data and transmits reliable data and status information to any (up to 3) upstream outstation not more than 1500 m away. This maximum distance is based on the reasoning that it is of no use to warn of things happening more than 1 minute driving away. Outstations are placed alongside the road near the gantries where they control the variable message signs. These outstations also have a microprocessor to process the detector data of up to 3 downstream detector sites, to transmit the results on request to a central computer and to control and monitor the overhead fibre optic secret signs.

These signs can show the advisory maximum speeds of 30, 50, 60, 70, 80 or 90 km/h, a red cross or a left or right pointing, slanting, arrow.

The central computer reigning over up to 375 outstations collects traffic data every 4 seconds and system monitoring data, like the legends actually shown, lamp failures and outstation status, every 20 seconds. The traffic data are fed into an AID algorithm which, when there is a queue, will fully automatically within half a minute have a slow down sequence of 70 and 50 out on the gantries. The legends are commanded by the centre to the outstations. In case the outstations lose contact with the centre, they will run their own internal AID program and decide whether or not to show a 70 or 50. The gantries (and outstations) are spaced about 750 metres on average. Not closer than sign readability distance (300 m) and not more than 1500 m.

Data transmission from detector stations to outstations and between outstations and the control centre is via standard (copper) transmission cables. For OS - CC communication one command and one answering party line is used for a set of up to 31 outstations. Transmission scheduling depends on the information to be handled and is taken care of by the centre.

Apart from fully automatic AID measures, the system can also be used by motorway maintenance personnel and by the police for lane closures and speed measures. MCSS comprises five (remote) workstations for operators. A cold standby second central computer facility can be used for research, data collection and as an MCSS simulator for training and testing.

The signalling philosophy is that correct and straightforward information should be given to tell the drivers what to do, instead of alerting or warning drivers and leaving it up to them how to cope with the traffic problem. This has led to the complete control and checking of all legends set up. This also requires a very high degree of reliability. The system layout reflects this; instead of falling back on duplication, control is distributed over the various system elements in a way which ensures that whichever element breaks down, the system will continue to operate safely. This is the graceful degradation philosophy mentioned before. The taking over of queue protection by locally operating outstations which have lost contact with the centre is one example. The overlapping of surveillance areas by outstations, using information from the same detector station close enough to be of interest, is another. Third, for OS - CC communications two cables are used, one for the even-numbered outstations, one for the odd-numbered ones. Even when a cable fails, a fair degree of surveillance and signalling is possible.

Credibility requires that the information given is correct in time and space. This is the reason why detector and gantry spacing is, when compared with other systems, close.

Various options are hardly used, simply because the sites now being covered by the 400 outstations do not need them. Such options are ramp metering and rerouteing. Although it is possible to show additional information, experience with explanatory, or enhanced, signs has led to the conclusion that they are normally not worth the cost.

INTEGRATION OF URBAN NETWORK AND MOTORWAY CONTROL

Traffic management for urban networks differs substantially from management of motorways. Normally there is no need at all to let the two interfere to get a better overall result. The exception is the oversaturation case. Both systems then might want to use a metering mechanism, which can affect traffic leaving one system to enter the other system. One can study the consequences by studying the two networks (urban and motorway) separately. Good overall study tools are not available. Steps in that direction are taken by the integration of TRANSYT and FREQ.

Also little practical experience is available. The IMIS

(Integrated Motorist Information System) in the USA comprises a complete network, but it operates as a combination of separate systems. The same is the case in the Glasgow system called CITRAC. Concluding, one can say that the lack of one all-encompassing system is no good excuse for not having good overall control. When traffic engineers know how to handle their motorways and their urban networks they can, forgetting personal, historical, political and administrative antagonisms, easily join hands and create a good interaction between their schemes to help each other fight oversaturation problems.

(6)

(7)

REFERENCES

Albrecht, H., Philipps, P. (1977). Ein Programmsystem zur verkehrsabhaengigen Signalsteuerung nach dem Verfahren der Signalplanbildung, BVM Bonn, Heft 240.

Baang, K. L. (1976) Optimal Control of Isolated Traffic Signals. Traffic Engineering and Control. July 1976.

Basispecificatie, (1980). Rijkswaterstaat TXR. The Hague.

Bretherton, R. D. Five Methods of Changing Fixed Time Traffic Signals. TRRL Report LR 879.

EUCO-COST 30 BIS (1982) The International Public Demonstration on Electronic Traffic Aids on Major Roads. European Comm. XII/952/82-EN Brussels.

Gartner, N. H. (1982) Development and Testing of a Demand Responsive Strategy for Traffic Signal Control. American Control Conference. Washington DC.

Hall, M. D., Van Vliet, D. and Willumsen, L. G. (1980) SATURN: a simulation-assignment model for the evaluation of traffic management schemes. Traffic Engineering and Control, Vol.21, 168-176.

Hilgers, C. J. (1980) A Method for Calculating a Group of Algorithms for Automatic Incident Detection. Journal A, Volume 21, No.3.

Hunt, P. B. et al SCOOT (1981) TRRL Laboratory Report 1014. Crowthorne.

Jovanis, P. P. et al (1978) FREQ6PE. ITTE. University of California Report No. UCB-ITS-RR-78-9.

Klijnhout, J. J. (1984) Motorway Control and Signalling - the test of time. Traffic Engineering and Control. Vol.25 No.4 April.

Kuhne, R. et al (1984) MM-Wave Based Traffic Flow Detection and Control Using a Stochastic Continuum Approach of Speed Distribution. Symposium Land Vehicle Navigation. Deutsche Gesellschaft fuer Ortung und Navigation, Muenster, Germany.

Leonard, D. R., Tough, J. B. and Baguley, P. C. (1978) CONTRAM: a traffic assignment model for predicting flows and queues during peak periods. TRRL LR 841.

Miller, A. J. (1965) A Computer Control System for Traffic Networks. Proc. Sec. Int. Symp. on the Theory of Traffic Flow. OECD. Paris.

Middleham, F. (1983) Flexsyt. An Ultimate Study Tool in Traffic Signal Systems. Transportation. An International Overview. ITE 53rd Meeting, August 1983. London.

OECD Road Research (1981) Traffic Control in Saturated Conditions. OECD Paris.

SCAT System (1979) Department of Motor Transport. New South Wales, 2nd Edition.

Signalhandboken (1982) TFK-rapport 1982: 3. Stockholm.

SIGSIM (1980) Manual. VBB. Stockholm.

Texier, P. Y. (1980) Role de la regulation dans le traitement de la saturation. TEC No.39. March-April, Paris, 1980.

Vincent, R. (1986) Self-optimising Traffic Signal Control using Microprocessors. TRRL "MOVA" Strategy for Isolated Intersections. IEE 2nd Int. Conf. "Road Traffic Control". London.

EDITORS' NOTES

(1) Although the induction loop is the predominant form of vehicle sensor used in Europe, there are a number of alternative technologies available. For a review of these see Chapter 2.

(2) A number of technical terms are used in this Chapter. We define some of them as follows:

<u>Phase or</u>
<u>Group</u> : a set of traffic streams receiving identical signal indications.

Stage : the interval in which a particular set of groups has green and the signals are not in the process of changing.

Cycle : a sequence of stages which is repeated. Cycle time is the duration of this sequence. If the sequence changes, a more complicated definition is required.

Extension: stage extension.

(3) For further details, see Robertson (1969) "TRANSYT: A traffic network study tool", Road Research Laboratory Report LR253.

(4) A recent study of the ageing of fixed time plans has thrown more light on this topic (see Bell and Bretherton's article entitled "Ageing of fixed-time traffic signal plans" in the Proceedings of the Second International Conference on Road Traffic Control, Institution of Electrical Engineers, London, April 1986).

(5) For a discussion of the media used to transmit messages to drivers on motorways, see Chapter 14.

(6) IMIS (Integrated Motorist Information System) is described in Saxton's paper entitled "Traffic systems in the United States; present and future developments" in the Proceedings of the International Seminar on Electronics and Traffic on Major Roads, Paris, June 1985.

(7) CITRAC (Centrally Integrated TRAffic Control) is described in Mowatt and Young's paper entitled "CITRAC - the first five years" in Traffic Engineering and Control, 251-259.

Information Technology Applications in Transport, pp. 141–164
P. Bonsall, M. Bell (Editors)

Chapter 7

COMPUTER SYSTEMS FOR THE PLANNING, MONITORING AND CONTROL OF RAIL OPERATIONS

John Holt
British Rail, UK

OVERVIEW OF BRITISH RAIL OPERATIONS SYSTEMS

This chapter describes the computer systems used by British Rail (BR) for the planning, monitoring and control of its rail operations. BR's general management systems (eg marketing, accounting, and stores control) and other specialist systems (eg signal co ordination, seat reservations, computer aided design, and maintenance control) are not described here.

BR's operations systems have evolved over the last decade and are still under development. The major systems involve very large databases which are housed in an IBM mainframe complex. The mainframe is linked to interactive outstation terminals by an extensive data transmission network (1,200 route miles) and by a High Speed Data Ring (HSDR) which provides for 2,000 on-line terminals throughout BR. Area level micro computer networks are being linked to the HSDR and, for some applications, this enables input or editing of a purely local character to be built into the national systems. (1)

This chapter will consider, in turn, train service planning, scheduling (known on the railways as diagramming), rostering, monitoring and control.

TRAIN SERVICE PLANNING

Introduction

Train service planning involves the timing and pathing of individual train services, their combination in BR's Train Service Database (TSDB) and production of timetables and other service information. A summary of the systems involved is shown in figure 1.

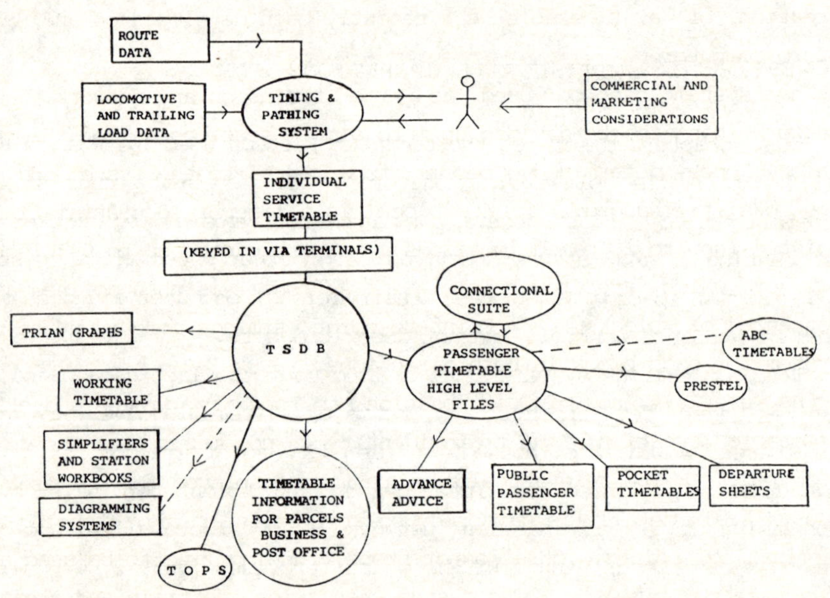

FIG 1: TRAIN SERVICE PLANNING SYSTEMS

The Timing and Pathing System

A computer program for the calculation of the point-to-point times for trains has been in regular use for many years. It uses three basic types of information:

(a) Route data - timing point locations, speed restrictions and gradients

(b) Locomotive data - eg gross tractive effort/speed, resistance/speed, fuel consumption/power and the mass and length of the locomotive

(c) Trailing load data - number of vehicles, mass, length and resistance/speed.

It applies the appropriate formulae to this data and produces a list of timings for the route and train service in question. This (2) is a mainframe batch process.

A system to enable this type of information to be used in the creation of a timetable has recently been developed on an IBM XT micro computer.

The system can be used to time trains on any network. The network has to be divided up into simple sections called 'Blocks' which, once determined, become fundamental to the system and are not easily changed. A 'block' is a line of sequential timing points (up to 24 named locations). There can be up to 64'blocks' all of which are reversible. A route is then defined as a sequence of blocks. Point-to-point timings are input to the system for each block in both directions.

The user is now in a position to specify trains and start timetable creation. A train identity is now input with the route required. The relevant blocks are then displayed for the user to input train detail (ie days run, power type, starting, stopping and terminating points, duration of stops, running line to be used and any pathing time). The system then times the train and creates the equivalent of a timetable column for the user to examine. This can then be amended, re-timed etc. Many train services can be created in this way very quickly.

If, when the user has specified and timed a whole service, he wishes to print out a timetable he defines the station bank (or route) of his timetable, and the system then sorts the data into proper order and prints the timetable.

Further development work is currently in hand to enable timings so created to be directly input to the TSDB and to identify and highlight conflicts between trains. At present this latter aspect is overcome by the use of experienced staff.

The Train Service Database (TSDB)

The TSDB is a production system which has been in use since 1979 and provides the base data for timetable production and other associated systems. It has been subject to an evolutionary process which will continue as the associated applications increase in number, and technology changes. The TSDB stores information on timetabled train services for the whole of British Rail. It includes the permanent timetable and monthly updates but not traffic planned at very short notice. Neither does it include full details of resource allocation.

It is a true database system using the database management system ADABAS Version 4.1 which means that it can be accessed relatively quickly for a wide variety of uses. There are approximately 50,000 trains on the database with 650,000 events (passing times, stopping times, etc.) covering 6,500 locations. The size of the database is approximately 130 Megabytes, basic train data taking up approximately 60 Megabytes. The data is currently input via Datapads (devices for recognising handprint) which are connected to a local mini computer linked by direct line to the mainframe computer. This system exists at all six Train Planning Offices. It is however being replaced with simple non-intelligent terminals linked via the HSDR to the mainframe computer. This will allow interrogation of all files for which the particular individual/terminal or office is allowed access, in addition to straight data input or file amendment/deletion.

The system is based on a 'plan year' concept whereby the current timetable is held as Plan Year 1 and the build-up situation for next year is held as Plan Year 2. The build-up of Plan Year 2 for the next timetable takes place progressively from May until December. In the second week of December the data is 'frozen' for onward processing to provide timetables etc. While it is possible to amend the database up until the 'roll-over' which takes place in mid April (ie approximately four weeks in advance of a timetable coming into operation) this is not desirable because some information will already have been taken from the database to produce documents which will have to be amended separately.

The plan year concept has proved cumbersome and so a more flexible system is being developed for implementation by 1987. TSDB will then become a continuous database with all items date labelled.

The TSDB supplies the following systems/functions:

Simplifiers. At many locations within BR local staff produce extracts for their own use of timetable information etc.; these are referred to as simplifiers. It is now possible to produce many of these from TSDB via laser printing direct from the mainframe.

Particularly significant in this respect are Signal Box Simplifiers which can be quite complex and have slightly different requirements in different areas.

Station Working Books. Major stations have a Station Working Booklet produced. Because this contains more information of a purely local character the production process is somewhat different. The train service data required is extracted from TSDB on the mainframe and then down-loaded on to a floppy disc. This is then edited/manipulated by the local manager's staff on their IBM XT micro computer. The disc is then sent to an outside contractor for phototypesetting and printing.

Train Graphs. Within train planning offices time/distance graphs depicting the planned running of trains are basic planning tools. They are also used to examine the best timing for special or new services (both freight and passenger). A facility has therefore been created to automatically plot selected train services held in the TSDB on a CalComp 4 Colour Plotter.

Working Timetables. The working timetables are produced from the train service database in a similar way to the public timetable which is explained later. The main differences are that both Freight and Passenger trains are included, connections are not shown and passing times at juctions, signal-boxes etc. are included. The publication is a whole series of separate books generally containing only 2 or 3 tables, distributed in widely varying numbers. The final output of the computer system is a drive tape for a phototypesetter.

The Master Timetable System (MTS). MTS, which became operational in 1984, is unique to the Southern Region of BR and is part of the new signalling scheme recently installed, its purpose is automatically to report trains running late (outside predetermined limits) and to set platform indicators. For this purpose it clearly needs to store timetable information which, were it not for TSDB, would require a huge manual input. Schedules are extracted from TSDB along with additional parameters such as Stopping Pattern Codes which have to be checked against a special set of fixed data. The transfer of data takes place annually with a weekly update.

Passenger Timetables. The relevant trains are extracted from TSDB and allocated to each of the tables in the timetable alongside the station banks previously fed into the computer as fixed data. These are then sorted into their correct order and notation generated relevant to the "black type" (through) trains. At this stage the master file contains all the information required in each timetable with the exception of connections.

The connectional suite is based on a relatively straightforward algorithm which uses fixed data defining acceptable routes and interchange times (by location) to calculate the earliest connection to the places defined in the main timetable. After the connectional suite has run, the information is entered on to the master file to become the 'Passenger Timetable High Level File'. A number of computer printouts are produced which alert the timetable compiler to certain situations such as "just missed" connections or those too tight or too long. The compiler can then make any adjustment by editing the file to be printed before it is passed to the phototypesetting process.

Outputs include; Advance Advice sheets; the BR passenger timetable itself; station departure posters; pocket timetables and information for British Telecom's Viewdata system - PRESTEL and for ABC timetables.

Advance Advice sheets can be prepared at any time by a laser printer and are distributed to Area Managers, Travel Centres etc.

The BR Passenger Timetable requires a series of programs to be run which paginate the timetable, insert typesetting commands and ultimately produce a magnetic tape to drive a phototypesetter (Typically VIDEOCOMP S70E or APS5) which in turn produces either a film or bromide at A4 size. These are sent for printing in both A4 and A5 size copies for internal use and public sale. This system has reduced both typesetting costs and the production timescale.

Station Departure Posters at each station present the services to selected destinations in alphabetic order. The data is loaded by the mainframe to diskettes suitable for micro computers in the Area Managers offices. (Subsequently this may be directly down-loaded.) The diskettes are edited and extra local information added as necessary on either SIRIUS or IBM XT/PC micro computers. These are then returned to a mainframe for collation and addition of typesetting commands prior to going to a printers for final production.

Pocket Timetables are produced in a similar manner to Station Departure Sheets. However more fixed data is necessary to define the front cover and the layout preferred. Also the notes are inserted at different levels within the organisation (ie Nationally, Regionally, and Locally). The user requires a terminal linked via the HSDR to the mainframe computer. He creates the fixed data interactively, requests the extraction run and can then edit the data. The display simulates the final publication as far as possible. On completion it is released for typesetting and subsequent printing.

PRESTEL contains timetable information used primarily by travel agents. Customised software manipulates the data into a form which can be displayed on Prestel screens. This system has enabled in excess of 10,000 frames to be created very quickly with little manual effort at each change of timetable.

ABC Timetables are supplied with some timetable information via magnetic tape for direct use in their own system. (3)

Mails and Parcels Timetables. Timetables for BR's express service, Red Star, are produced in a similar manner to Pocket Timetables from TSDB. It is envisaged that the Parcels Business Machines, which require timetable information electronically, will also be loaded from TSDB. The Post Office are provided with special printouts of selected trains for use in their mail delivery system. Consideration is being given to direct transmission of data to their computer system.

Diagramming Systems. There are separate systems for scheduling crews, locomotives and other rolling stock, all of which require data from TSDB and are described in a later section of this chapter. Not all of them have yet been given a direct interface with TSDB.

The TOPS System. Last, but by no means least of the uses of TSDB data, is BR's Total Operations Processing System (TOPS) also described in a later section of this chapter. Train schedules for TOPS are extracted from TSDB at roll over time and then updated in respect of any weekly alterations.

(4) DIAGRAMMING AND ROSTERING

Linkages

Figure 2 summarises the inter-relationship of BR's current diagramming and rostering systems, all of which take data from TSDB and feed into TOPS. The link with TOPS is shown via the, as yet incomplete, Diagram Input and Automatic Distribution System (DIADS) which is being developed to facilitate the clerical processes of amending and distributing diagrams, and to store the information in a manner suitable for use by other computer systems.

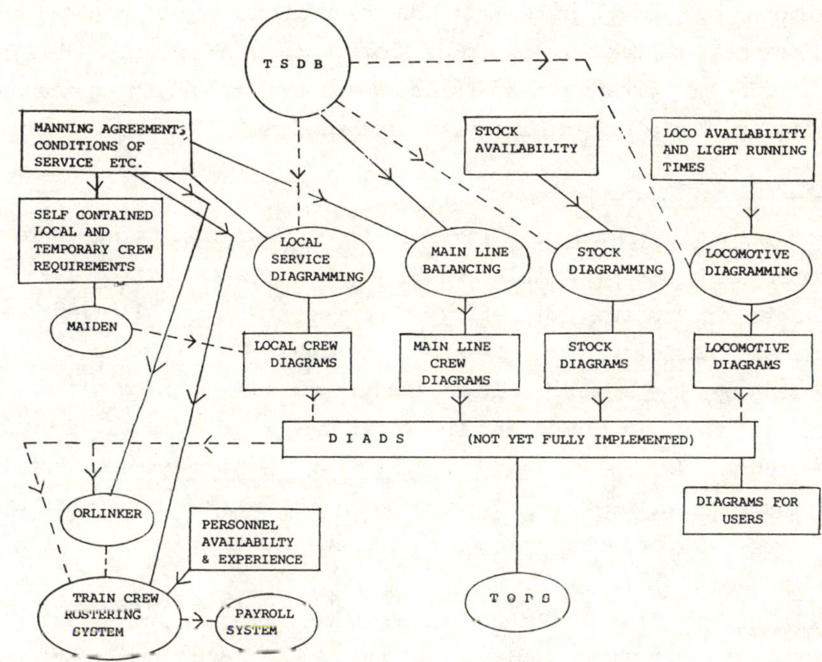

FIG 2: DIAGRAMMING AND ROSTERING SYSTEMS

The completed system will allow Crew, Stock or Locomotive Diagrams to be input to the mainframe computer through remote terminals displaying formatted screens calling for the relevant data. The system has validations which are carried out before the diagram is stored within the system. Diagram information can then be either passed to other computer systems, edited/amended on the terminals in the train planning offices or transmitted to the Area offices or Depots and printed out locally where they are required. Any authorised terminal with access to the TOPS network will be able to receive diagrams.

Once the Diagram data is universally available systems will be designed to automatically produce various analyses such as the productive content of diagrams and to distribute them to relevant offices.

It is planned to eventually use the diagram data within the TOPS system for the control of train crews. Day to day alterations would be entered in the system. Managers and supervisors would therefore have full information of the availability of men and the requirements of the diagrams plus spare time in diagrams. This is considered to offer considerable scope for exploiting unbalanced working and for other productivity improvements.

Vehicle Diagramming

Locomotive Diagramming. The locomotive diagramming program is a computer aid not a total system. The objective is to use the available locomotives as efficiently as possible allowing for practical considerations. The efficiency of the result is measured by the amount of light running (ie time running without a train in tow) involved and the total number of locomotives used. The program is interactive and relies on the user knowing what is practically feasible. It runs on an IBM mainframe computer and the user operates it through a remote Time Share Option (TSO) terminal and printer.

The basic 'tool' which the computer program uses is the 'ASSIGNMENT ALGORITHM' which, on the basis of the 'cost' of connecting together any given pair of trains, calculates the cost of diagramming all the trains together or of diagramming any subset of these trains.

The program input is: a timetable of trains and their locomotive requirements; the types and numbers of locomotives available; and light running times between each pair of terminals (stations, yards, etc.).

The program initially calculates the overall peak requirement, ie the theoretical minimum number of locomotives required to cover the trains. The user must then decide how many locomotives in total he is prepared to use. To assist this the computer program produces a list of light running versus number of locomotives for all the trains together. The user can now start to diagram individual locomotive types, using this background knowledge by inputting the number of locomotives of the selected type he wishes to use, plus other information to influence the kind of solution to be obtained. When the diagrams are produced the computer program indicates the cost (in light running terms) of the solution and any additional cost which is implied in diagramming the remaining trains.

If the diagrams produced are not acceptable the user may rework the locomotive type giving additional information to the computer program to give a more acceptable result. This process is continued for all the locomotive types until the user has a complete set of satisfactory diagrams. The system is flexible in that any previously diagrammed locomotive type may be brought back and re-diagrammed if necessary at some later stage in the process.

The final outputs are in the form of bar charts plus lists of light running, locomotives used and an index indicating by train headcode the locomotive type on which each train is diagrammed.

At the present time this system operates in stand alone mode. Interfaces to the TSDB and other systems are planned for the future.

Rolling Stock Diagramming. Locomotive hauled coaching stock is currently diagrammed manually and then entered into the computer via DIADS. The diagrams are subsequently used within the TOPS system for the Passenger Operations Information System referred to later. Freight wagons are not normally diagrammed.

Train Crew Diagrams

Main Line Train Crews. The computer system used to create practical and economic diagrams for train crews on longer distance train working is known as Main Line Balancing (MLB). Such a diagram typically consists of an outward journey, followed by a meal break, followed by a return journey to their own depot, but can involve working 4 different trains. The system operates under TSO using remote terminals to the IBM mainframe computer.

Train crews have to be consulted regarding their diagrammed working. For this reason the system has to be much more flexible in its application. This has led to a system which can also evaluate the effects of changes in working conditions, train services, methods of working, etc.

MLB requires a large amount of fixed data on manning agreements and conditions of service (permissible hours, mileage limitations, walking/signing on allowances etc.). The process commences by extracting the train services to be covered from the TSDB for a given route or section of route. The diagramming clerk then proceeds to work inter-actively on the terminal using a railway oriented command language.

Initially the system assigns trains to train crews using an algorithm which minimises both the number of crews required and the unproductive time at the outward end of the journey.

A range of commands exist to deal with assignments involving up to 6 locations with crews travelling passenger between relief points when necessary. Triangular balances can also be achieved. The effects of limiting the number of crews available at specific locations, restricting the sequence of manning specific trains etc. can also be explored. A large network can be progressively covered using these commands until a set of diagrams is produced which are both economic and practical. The results can be output to printer, or compared within the system with alternatives previously considered.

Where used for long term planning or strategic purposes, the manning agreements, allowances, conditions of service, can be changed and the effects compared.

The diagrams prepared in this manner have at present to be manually edited to incorporate short make-up type jobs, and allow for daily variations. Developments to transfer the 'skeleton' diagrams to the DIADS system for editing and distribution are currently in hand.

Local Services Train Crews. The objectives are similar to those of MLB, however the long string of activities typical of intensively worked suburban areas generates far too many options. An aid to diagram creation is consequently under development which displays in bar chart form the traction/multiple-unit schedules, and then allows the planner to inter-actively cover these schedules with crews. As this is progressively carried out the proposed crew diagrams are validated against National Agreements. A technique to highlight the peak demand (defined as the list of train services, none of which could possibly be operated by the same crew) is also being explored as an additional aid to ensure minimisation of the number of crews required. This system when operational will interface with DIADS.

Micro-computer Generated Local Service Diagrams (MAIDEN). In some local areas of B.R. small numbers of locomotive and train crew resources are allocated to cover self-contained work which can also be subject to change from day to day. In order that diagrams, which conform to National Agreements, can be produced for the train crews, a micro computer (IBM XT) system for local use has been produced. The menu driven system is designed for users unfamiliar with computers but with sound local railway operating knowledge. Eventually this will be linked to the mainframe systems.

If an inconsistent or illegal activity or location is requested, the program will intercept it and reject it. The program does not at this stage attempt to optimise. All that is intended is an aid to local staff producing diagrams.

The following "fixed data" is required: a comprehensive table of running times; a list of terminal times for each location in the area, allowing for every possible activity (eg in loaded - out loaded or in light - out loaded); details of general constraints (eg route restrictions, limits of route knowledge for crews, locations where a break can be taken); and the allowances for train and locomotive preparation, signing-on and signing-off.

The advantages of the system are that accurate diagrams can be produced very quickly and printed out ready for immediate use in a local situation.

Traincrew Rostering

Design of Train Crew 'Links' by computer (ORLINKER). For convenience in allocating men, each train crew depot has its work grouped into 'Links'. These limit the route and traction knowledge required, and may group similar types of work. A 'Link' defines a sequence of working diagrams which, if followed by a man, will enable him to have acceptable rest periods, average 39 hours a week over a period of 8 weeks, and be fully utilised without incurring excessive amounts of route learning. The process of allocating men to 'Links', and allowing for holidays, sickness, etc., is known as rostering.

Each time new diagrams are issued as a result of changes in the train service the 'Links' may need alteration and this can be a very complex process. A computer program (ORLINKER) has therefore been designed to assist in this task. It is used also to evaluate the cost effect of following different strategies and policies.

ORLINKER uses a TSO terminal. Initially it takes the crew diagrams produced by the train planning staff and produces a set of links which cover them, together with other summary information (eg route/traction learning costs, average earnings, etc.). To produce these links the user has to provide information about the agreements, strategies etc. that are to be used, or evaluated. The system then attempts to produce links which use the minimum number of men to cover the fixed workload, (ie fewest possible lines (weeks) of work and minimum route learning) and to choose the sign-on time, rest days, etc. of the spare men* so that the proportion of days on which the spare men are used to cover absence is a maximum. (*'Spare' men is the term used for those men used to cover sickness, holidays, other forms of absence and special traffic requirements.)

This system functions in a stand-alone mode at present. There are plans however to interface it with DIADS for diagram input, and with the rostering package to save clerical time and effort.

<u>The Rostering System</u>. This system assists the production of daily and weekly rosters, daily supplementary sheets for train crews and provides management information.

The Rostering System is designed to work on an IBM XT, and an Epson MX or FX 100 printer in stand-alone or networked systems. The system uses the rostering programs, the depot's roster and rostering rules, details of men's pay numbers and their route and traction knowledge.

On a daily (or weekly) basis, the operator keys in details of the men who are absent, resuming duty, training or on other duties, details of specials, cancellations and other alterations to the booked services. The system will then generate time-order lists of uncovered jobs and spare men according to the rostering rules entered into the machine. These lists of jobs and men can be fully checked by the machine to ensure that men have the requisite route and traction knowledge for the job and are within the two hours movement allowed in the National Agreements for daily rostering.

From the time-order lists, the roster clerk chooses who should be allocated to which job, according to the practice at the depot, and types pairs of index numbers into the machine, thus making the allocations. When the allocations are complete, the machine prints out daily rosters, appearance sheets or alteration sheets, and "spare" lists, etc. The machine will produce several copies of the printouts as required.

Many other different types of listing are possible, both daily and weekly. There are plans to subsequently interface this system with the Direct Data Entry system for the National Payroll System.

TOTAL OPERATIONS PROCESSING SYSTEM (TOPS)

Introduction

The TOPS system (summarised in figure 3) enables the resources engaged in train movement to be monitored and controlled for optimum achievement against traffic requirements.

The current status and location of locomotives, locomotive hauled passenger and freight rolling stock, can be input from a multiplicity of reporting points throughout British Rail. This information is immediately accessible by users. In most cases it is output automatically to those who will inevitably need it to carry out their work.

At higher management levels, operations can be monitored and priorities can be decided in times of shortage of resources.

Resources can also be effectively balanced between areas of surplus and deficiency, and between traffic operation and maintenance.

In addition, statistical information is available from the off-line part of TOPS known as TARDIS (Tops Ancillary Retrospective Data Information System), and freight customers can have terminals with direct access to the TOPS computer to monitor the progress of their own traffics.

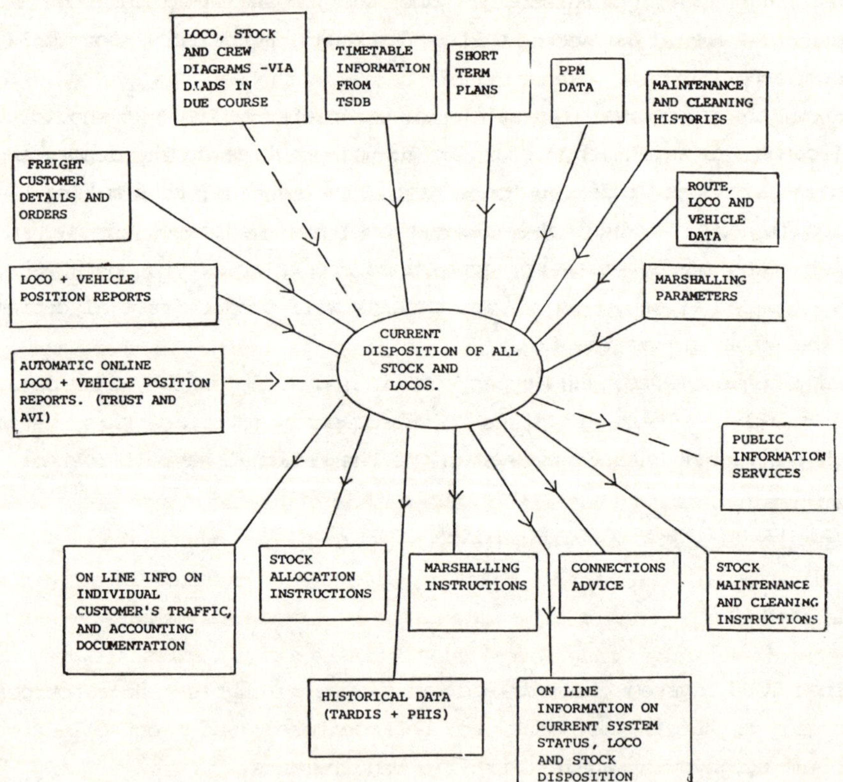

FIG 3: TOTAL OPERATIONS PROCESSING SYSTEM (TOPS) INCORPORATING POIS

Specific Applications Within TOPS

Freight Wagon Distribution. The system is semi-automatic in that the distributor is only responsible for ensuring that the demands for wagons are being correctly reflected within the computer's on line files, also that the average movement and loading time for empty wagons at each location is unchanged. (This information is supported by weekly off line statistics.) However, the system has a large number of economic options for wagon supply to each demand point and the computer destines wagons automatically everytime they are reported unloaded. These options are structured to meet real demand, anticipated demand and storage of longer term surplus of wagons. Where commercial priorities override operating economics the distributor can override the normal priorities executing his decision through the system. The system is self monitoring in that any indiscipline in the field is immediately reported to the distributor. The distribution of a fleet of 81,000 wagons is monitored 24 hours a day and controlled by 10 staff. Subsidiary systems are built in to control wagons in assigned working and special wagons, whose distribution is individually controlled.

The system improved wagon availability by 37% in the first two years after implementation.

Marshalling Yard Management. The system combines routeing information with details of the method of marshalling used in each yard to produce, in advance of trains arriving, shunt and cut lists for working the yard. Similarly lists are produced in advance of departure to enable the planning of train loadings, ensuring connections are made and that operating safety rules are complied with.

Cripple Wagon Control and Planned Preventive Maintenance. Using the same basic logic as for Wagon Distribution, when a wagon becomes defective it is automatically destined to the nearest grading location for inspection. The computer then destines it to a workshop which has the facilities and capacity to undertake the work involved.

Planned Preventive Maintenance (PPM) is carried out on the newer generation of wagons and certain wagons in assigned working. The computer files hold the specified period between PPM's and automatically allocates them to Maintenance Depots when they become due.

On-line Enquiry Facilities. The exact state (wagon by wagon, and siding by siding) of all yards is available to anyone with a TOPS connected terminal. Information as to permitted load, brake force requirements, route availability restrictions etc. for every freight train and its composition, locomotive and location are similarly available to all and can be captured to measure productivity. Connectional arrangements for important traffics can be anticipated in advance by supervisory control centres. The current state of vehicles in relation to servicing and maintenance can be readily checked.

Centralised Motive Power Control. TOPS printouts provide information to the Controller as to the location and status of every locomotive in his area of jurisdiction, plus details of those locomotives en route to his area and those allocated to arrive in the future but not yet en route. It also provides details of when locomotives are next due the various categories of maintenance or workshop attention. Similarly, it automatically advises everyone concerned as to each Controller's decision regarding allocation of locomotives. All these tasks had previously to be carried out manually.

The Controller plans and allocates each locomotive not only for its present task but two more tasks in advance. Thus he makes his dispositions to trains up to 8 hours in advance and assigns locomotives for maintenance up to 7 days ahead. The Controller can therefore anticipate and iron out peak demands, respond to stated commercial priorities, and anticipate and remedy emerging local shortages. The benefits of the new system were immediate in that as well as abolishing some 240 Controllers' positions, the fleet availability immediately improved by some 2%.

<u>Freight Consignment Processing</u>. In 1975 it was decided to eliminate the use of weighbills for freight traffic as proof of delivery and instead use the TOPS system for consigning and charging information. The system is known as 'Consigned Through TOPS' (CTT).

The basic concept is very simple; a flow file in TOPS gives a unique identification number to every flow of traffic on British Rail. Customers are issued with copies of this file in the form of cards which record all the basic detail including the charging information relative to their flows of traffic.

When a wagon is reported to TOPS as having been loaded all that is necessary to report is the wagon number, flow number and content weight of the wagon. The computer then provides all the necessary operating information to staff in the field and, at the same time, generates a record to the accounting computer of the consignment and the rate folio number, so automatically processing the customer's account.

Some customers have been equipped with their own TOPS terminals and, in fact, make direct input of wagon release information (loading and unloading information) to the central processor, or use the National telex system for input to TOPS. This causes unsolicited messages to be output at the Area Freight Centres thus giving the necessary instructions for movement.

<u>Passenger Operations Information</u>. The TOPS system was originally oriented only to freight operations but now also embraces some passenger operations, with a system known as POIS.

Locomotive hauled and H.S.T. passenger trains are planned and operated in sets. Reporting is basically of locomotive and set number, with the computer translating the information into individual vehicle numbers and outputting fully detailed train reports. All the coaching stock diagrams are held on file to enable the system to compare the actual with planned operation and output messages to controls, stations and depots to correct any malfunction of the plan.

All work done, maintenance, cleaning, and train running is reported to the system via hand-held terminals connected on a dial up system or by VDU terminals. The programs simply require a yes/no input for reporting work done and a train identity set number and time for movement reports. This enables the same terminal to be used by maintenance, carriage cleaning, or station supervisors by simply asking for the appropriate program. Output response can be sent via the hand-held terminal to a simple associated printer.

Under POIS all coaching stock is nationally controlled. Any requirements for special trains are input to the system by the Regional Train Planning offices before accepting a commitment, and an allocation is determined by the central distributor. On introducing POIS, savings resulted from a considerable reduction in clerical staffing, reduced empty coaching stock mileage and a reduction in the fleet of vehicles.

The computer also accumulates the mileage and time in service for every vehicle and train set, and it is therefore possible to apply a national standard for carriage maintenance and shopping. Automatic outputs from the system enable depot workloads to be planned in advance and prevent under and over maintenance taking place. Similarly, on the basis of production targets, the system estimates the time and date when vehicles in workshops should be available for service. Work-done reports record the maintenance history of each vehicle.

The system automatically draws the attention of depots to sets which are due for heavy cleaning and requires a positive input to confirm the work has been done or, in emergency, that it has not been done before allowing further movement to take place.

Because the information in respect of any changes in train set formation is input to TOPS from the carriage sidings it is possible to advise reservation offices in advance of any alterations in seating capacity. Any necessary emergency action can then be taken before the carriages are placed in the departure station. The facility exists to input a count of the number of seats occupied on the train when reporting arrival. This is separately analysed by computer for marketing purposes.

<u>Train Running System/TOPS (TRUST)</u>. At the same time as TOPS has been expanding to cover more trains, so the modernisation and automation of signalling has been progressing. It is now possible on some routes for the computerised train describer equipment in the signal box to report trains passing preselected points to the TOPS system. Thus actual running can be compared with the train plan and reports generated. This system which is currently being developed will come into use on the Euston - Manchester, Liverpool and Glasgow route in 1986.

The selection of locations at which reports will be passed from the signalling system to TOPS will be on average 15 minutes apart at class 1 speed, but this will be varied so as to be meaningful according to local circumstances.

A facility to allow the manual insertion of information on the cause of delay, with common causes codified, is also to be incorporated. Where delay is being caused to all trains, passing over a section of line (eg permanent way restriction), one input report will cover all the affected trains. Subsequent analysis will then be possible to assist management in taking remedial action.

The computer terminals to which reports may be distributed can be any terminal attached to the TOPS or National Teleprinter Network (NTN) systems. The selection of terminals requiring a train report will be automatically generated. To avoid the possibility of receiving terminals being swamped by reports of trains running slightly late, a delay factor will be built in to prevent automatic distribution of reports until the train is more than a certain number of minutes late. This number of minutes can vary for different classes of train and for each terminal.

In addition, it will be possible for any TOPS or NTN terminal to ask for the latest available train running information on any train. The information gathered by TOPS will also be used to generate punctuality statistics. (In the Passenger Historic Information System - PHIS). Further development work is in hand to enable this system to be interfaced with the Public Information services at the larger main line stations.

Automatic Vehicle Identification (AVI). The TOPS system, like many others, depends on data input which is costly in manpower terms and subject to the sort of reliability associated with manual input. The development of the AVI system is aimed at automating a major part of this. It depends on a vehicle mounted transponder into which is encoded the vehicle's number and a line side interrogator which reads the number as the vehicle passes (5)

The initial application under development (shown in figure 4) is designed for coal trains passing between collieries and power stations. It has four main components:

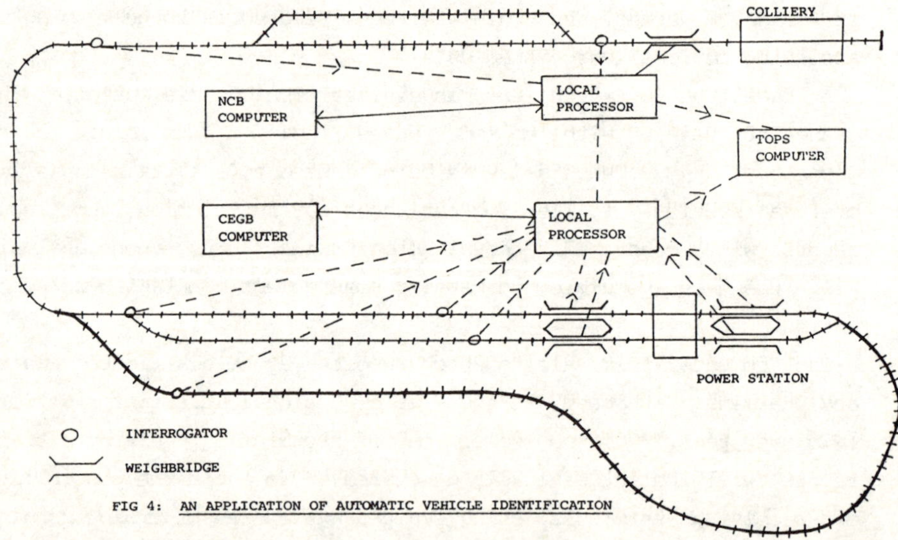

FIG 4: AN APPLICATION OF AUTOMATIC VEHICLE IDENTIFICATION

1. Vehicle mounted transponders The encapsulated transponders mounted under each vehicle are coded with the vehicle number, a check digit and an indication of the number of axles on the vehicle. When the vehicle passes an interrogation point, the transponder receives inducted current from the aerial system and uses this power to transmit the coded data to the interrogator.

2. Trackside Equipment Interrogators in lineside cabinets supply power to the vehicle transponders via an aerial on the track. They then read the transponder data, store the characters and

forward the data to a local micro processor. Data transmitted includes: vehicle identification, direction of travel, time, date and geographic location.

Aerials are necessary wherever data capture is required, (e.g. entrance, exit, weighbridge etc.) and are accurate up to speeds of 55 m.p.h. Axle counters sum the number of axles, give direction, and provide an independent source to indicate that a transponder interrogation should take place.

3. Micro Computer Standard hardware provides the following primary functions:

- To accept and store serial data from Transponder interrogators.
- To accept and record serial data from weighbridges and match it against appropriate wagon numbers received from interrogators.
- To format and process serial data.
- To accept customer details from data terminal/magnetic card reader or pre-programmed details from other devices.
- To report wagon numbers, consignment details, status change and arrival/departure times to British Rail's TOPS computer and to other computer systems as required.
- To store data in memory and produce a hard copy as required.
- To accept/transmit train bill data.
- To compare train billwagon numbers and weights with local weighbridge weights and advise disccrepancies.

4. Auto Dial Modems Modems with auto dial and auto answer facilities are an integral part of the system. Their bi-directional capabilities enable the micro computer to communicate and exchange data with other systems.

In principle the system records the wagon numbers and weights on entering and leaving the colliery, and again at the power station. This information is reported to TOPS for control and billing purposes. The system also reports to the customer's computer system for their accounting and control purposes. The whole system is designed for minimum operator involvement. Following this use of the system, its introduction into other spheres of railway operation is envisaged.

EDITORS' NOTES

(1) For a discussion of the role of locally networked microcomputers, see Chapters 10 and 12 in the contexts of the bus and travel industries respectively.

(2) It is interesting to contrast the amount of detail used to determine train running times with the approach adopted to solve the equivalent problem in the bus industry where the problem is primarily one of dependence on unpredictable traffic conditions rather than of susceptibility to vehicle characteristics.

(3) For a further discussion of innovative ways in which timetable information can be presented to intending passengers, see Chapter 13.

(4) For a discussion of the software used to solve analogous problems in the road freight and bus industries, see Chapters 9 and 10 respectively.

(5) AVI in the context of road traffic is discussed in detail in Chapter 3.

Information Technology Applications in Transport, pp. 165–189
P. Bonsall, M. Bell (Editors)

Chapter 8

AIR TRAFFIC CONTROL SYSTEMS

Nigel Ross
Software Sciences Ltd., UK

INTRODUCTION

The Development of Air Traffic Control

From the earliest days of aviation, pilots have been responsible for the safety of their aircraft and passengers. In the pioneering days with few aircraft and the desire to fly only in good weather, the main problem was preventing unscheduled contact with the ground. In good visibility at low speeds with open cockpits and few aircraft in existence, collision with other aircraft was not a major problem. Navigation at low altitudes was simple: follow features such as railway lines and, if lost, land and ask for directions.

The need to fly in all weathers to satisfy commercial demand, coupled with higher speeds and a higher density of aircraft, necessitated an effective Air Traffic Control (ATC) system which would provide a separation service whilst at the same time ensuring that traffic flow would be orderly and expeditious.

The main safety concerns confronting aviation which have been addressed by ATC are: collision with the ground and obstructions on the ground, collision with other aircraft, navigation and avoidance of undue delays.

The pattern that has evolved over the years is that aircraft flying in good visibility can travel in virtual freedom on a see-and-avoid principle with ATC providing an advisory service, except when taking off and landing at an aerodrome, where ATC instructions are mandatory. To link the routes most used by commercial aviation, corridors of airspace (termed airways) have been established. Airways are generally 10 nautical miles (nm) wide and ATC instructions are mandatory within them.

Additionally, aircraft have to be suitably equipped with the appropriate communications and navigation facilities and they must also pre-notify the ATC authority and receive permission or clearance to fly in the airways.

The ATC system should, ideally, provide capacity and facilities to allow aircraft to fly when and where the pilots wish. However, limitations on capacity make this impossible.

Airlines arrange schedules to meet the demands of their customers and this results in a high number of aircraft in the system at peak travelling times causing problems in both airspace and runway capacity. Another limiting factor is the need for airlines to minimise their costs by choosing the most economical route. Over large areas such as the North Atlantic, where set airways do not exist, routings are decided based on the disposition of high and low atmospheric pressure areas to endeavour to achieve a tail wind during the flight. Thus, whilst there may appear to be a vast amount of airspace only narrow bands will be selected by the aircrew. Demand on preferred altitude bands is also a problem as aircraft of similar type have the same economic cruising level.

Current Procedures

To illustrate how air traffic control works, we can follow a typical flight from its planning stage through take off, the en-route phase, and arrival at the destination airfield. The flight crew, having decided their destination, consult documents published by the national Air Traffic Control Agencies. These show details of the aerodromes which they are to use (including approach/departure procedures), the airways system, the relevant ATC authorities (together with the R/T frequencies on which they can be contacted),

(1)

and any special restrictions on the use of the airspace such as military danger areas. The crew will plan the route to join and leave the airways system at convenient points then fill in a standard form, termed a flight plan, which details the aircraft and its proposed journey. The flight plan is then sent to all the ATC authorities having jurisdiction over the planned route. It is the means by which a pilot advises ATC organisations of his intentions as regards destination, route and preferred altitude.

From the flight plan, the air traffic controller will be able to assess the loadings on the system and can either accept the aircraft's plan without modification or can issue a clearance subject to various alterations such as time of departure and the altitude to be flown. Prior to take-off, the flight crew will be given a time and altitude for the point at which the aircraft is to join the airways system.

The crew is responsible for the navigation of the aircraft and ensuring that the correct facilities are available for the safety of the flight. The crew will obtain a briefing on the weather along the route to decide if conditions are suitable for the flight. They will also have access to the Aeronautical Information Service which provides up-to-date information on the availability and serviceability of navigation and landing aids and details any hazards likely to be encountered whilst in passage.

Having received their clearance, the aircraft receives permission to start engines and is given instructions by the aerodrome ground movements controller to proceed to the runway in use. Further clearance is necessary for the aircraft to enter the runway and take-off. On take-off, the aircraft will navigate itself to the point at which it joins the airways and contacts the airways controller on the R/T frequency used by that section or sector of the airway.

The earliest form of control of airways still in use is termed procedural control. Aircraft are allocated an altitude or Flight Level (at 1000 ft intervals) and have to report their position at designated points, usually coinciding with a navigation aid (see below), along the airway. No other aircraft will be allowed to fly at the same flight level unless there is at least a 10 minute time interval between the aircraft.

Procedural control works well but has the disadvantage that the controller must rely on position reports from aircraft with no independent means of confirming the validity. Because of the 10 minute rule, designed to give a large safety factor, the airspace is used inefficiently causing reduced throughput and consequential delays. The availability of radar which provides controllers with a plan view of airspace enables aircraft to be independently monitored and allows the contingency margin to be reduced.

The general rule using radar is that aircraft should still be separated from each other by either 5 nm laterally or 1000 ft vertically. Compared with the separation standards for procedural control it can be seen that the introduction of radar has enabled a much higher utilisation of airway capacity, minimising delays as air transport has increased.

The last phase of our flight will be when the captain requests permission to commence descent for landing. This phase is probably the most complex as the flight will have to be guided to merge into the sequence of aircraft wanting to land. It is at this stage of the flight that landing capacity at an airport causes delays and an aircraft may have to fly a holding manoeuvre awaiting its turn to land. This is the dreaded 'stack' which builds up at peak times at busy airports and it is only when the aircraft has been given permission to continue for landing that control is handed over to an approach controller who is responsible for the final sequencing and landing phase.

Administration

The provision of an ATC service with its associated equipment is the responsibility of the national government working to guidelines laid down by the International Civil Aviation Organisation (ICAO)

Figure 1. The Scottish Air Traffic Control Centre showing the circular radar displays and the flight strip bays which may be used for procedural control. (Photograph Marconi Radar Systems Ltd)

which ensures interoperability and application of a set of minimum standards.

The provision of ATC services on a national basis is generally workable but there are areas where this division of responsibility could cause unnecessary delays. This is particularly true of Western Europe where there are a number of countries with small geographic areas having a high density of flights. The rapid handover of responsibility imposes severe co-ordination problems and as a solution a multi-national organisation known as the Eurocontrol Agency was formed to provide an en-route ATC service for the member nations (France, United Kingdom, Eire, Republic of Germany, Belgium, Netherlands and Luxembourg). Eurocontrol has set-up radar facilities and joint ATC Centres and administers the collection of fees from airlines on behalf of the member nations.

There is also a need to co-ordinate military and civilian air traffic and most nations follow the pattern of establishing joint organisations to plan and operate ATC services to enable maximum freedom of operation. Joint use of radars, communications and Air Traffic Control Centres is normal but the provision of ATC services at aerodrome or airport level is usually the sole responsibility of the operating authority. Responsibility for control of a flight will be transferred between sectors and over national boundaries. The handover of control is straightforward as the receiving controller will have had pre-notification of the flight via the flight plan and the handing over controller will have made direct telephone contact to pass details such as position and height.

Each national administration is free to implement its policies within the ICAO guidelines and consequently the standard of equipment and service is variable. The quality of service is generally related to the level of investment with factors such as aircraft density, terrain and weather determining the levels of equipment required. Most ATC equipment is expensive to purchase and maintain and, with highly complex ATC Centres, long planning, procurement and implementation cycles are normal. The following section reviews the current requirements and state of the art of equipment.

COMMUNICATIONS

Air/Ground/Air

Issuing clear and unambiguous instructions to aircraft is an essential part of Air Traffic Control. VHF Radio Telephony (R/T) is the universal communications medium allowing controllers and air crew to converse freely. As VHF R/T operates line of sight, a number of transmitter and receiver stations may be needed to provide adequate coverage of a large sector, particularly at the lower flight levels.

The provision of VHF coverage over areas of ocean or the less populated land areas is impractical and so HF, which has better propagation characteristics for longer range, is used instead. However the susceptibility of HF to atmospheric and ionospheric disturbance necessitates the involvement of a specialist in communications as the intermediary between the controller and aircraft.

The future trend in communications is to improve VHF coverage over densely used land areas and to examine alternatives to the HF system. The best proposition is the use of Satellites. However, the main drawback is one of expense. Satellite communication with aircraft was demonstrated in the 1960s and a programme known as Aerosat was examined by ICAO. Aerosat was intended as a dedicated aeronautical facility, but the finance was never forthcoming from the hard pressed airline industry and the programme was abandoned in the 1970s. With the introduction of the successful INMARSAT series to serve the needs of maritime communications, renewed interest in satellites has occurred. Provision has been made by INMARSAT for the inclusion of an aeronautical service in its next series of satellites scheduled for operation in 1988/89 and it is possible, though by no means certain, that an aeronautical service could start within a similar timescale. This would offer a relatively low-cost approach and the success of a pioneering service would renew interest in a dedicated aeronautical satellite. The demand for satellites may also be fuelled by a need for passenger communications and improved airline operations communications which cannot be met by the existing VHF and HF service.

There are problems with speech; a number of accidents allegedly having been caused by misinterpretation of instructions. To overcome this and to allow transmission of routine information to reduce loading on speech channels, digital data links have been investigated and are in limited use for the transmission of weather data and airlines operations data. It is to be expected that data links will emerge as an air traffic control medium, although there are a number of options for implementation which include Mode-S (see below).

Ground/Ground

A network of ground communications is necessary to provide transmission of flight plans, departure and arrival messages, meteorological data, radar data, links to transmitters and receivers and routine co-ordination requirements between Air Traffic Controllers at different units.

A combination of private wires and the public switched networks is used for most requirements with the transmission of flight plan data and Aeronautical Information being disseminated via a dedicated telegraphic system termed the Aeronautical Fixed Telecommunications Network (AFTN).

Operating at low speeds (100 baud with trunk routes of 2400 baud) AFTN systems operate on a star network principle. Message routeing is via switching centres which are now nearly all automated with computers having superseded the original manual operation. Transmission protocols are basic with no error correction, with reliance on message sequence numbering for detection of missing messages and initiation of retransmission requests. ICAO is planning a replacement for AFTN termed Common Integrated Data (2) Interchange Network (CIDIN) using X25 protocols to level 3, modified to allow for multi-broadcast to a number of addressees.

NAVIGATION AND SURVEILLANCE

Navigation

Navigation is the responsibility of the aircrew, but assistance can be provided by ATC.

The standard navigation aids in use are VHF omni-range (VOR), Distance Measuring equipment (DME) and Non Directional Beacons (NDB). NDBs radiate signals and aircraft can either home on them or carry direction finding apparatus to determine their position relative to an NDB. VOR provides the further facility of determining the bearing of an aircraft from the ground station but, contrary to its name, does not provide range which DME does. DME measures range on radar principles and VOR/DME are often co-located to provide a comprehensive navigation facility. In addition, modern passenger aircraft are equipped with inertial navigation systems of high accuracy, although to overcome human error in programming the systems, they can be updated and checked by VOR/DME inputs.

Future navigation systems are expected to be based on satellites. It is unlikely that the aeronautical community will be able to finance a dedicated satellite system for their sole use and it is probable that the US military navigation system NAVSTAR now being implemented will be the preferred system on grounds of cost and early availability.

(3)

Surveillance

Independent surveillance of aircraft allows controllers to monitor the navigation accuracy of aircraft allowing them to provide an effective anti-collision service and assisting full utilisation of airspace.

The universal surveillance aid in ATC is radar which improves on the inefficient procedural control method by allowing aircraft to fly closer together whilst maintaining high safety standards. There are two distinct types of radar in use termed primary radar and secondary surveillance radar (SSR).

Primary radar Primary radar, developed in the late 1930s, utilises a rotating antenna from which pulses of RF energy are transmitted. With each pulse, energy is reflected from an aircraft. The time interval between transmission and reception giving the aircraft's range. The position of aircraft are presented to the controller on a cathode ray tube on which map overlays can be displayed giving a two dimensional representation of geographic and airways layouts. In addition to aircraft being detected, radar returns are also received from precipitation, terrain and buildings, surface vehicular traffic, birds and atmospheric ionisation. These are collectively referred to as clutter and have the effect of masking the signals from aircraft. Various means of filtering out unwanted signals are available with recent advances in digital processing techniques offering marked improvements.

The most notable advance is Moving Target Detection (MTD). The purpose of MTD is to remove clutter which is stationary (terrain, buildings) and that which is slow moving (weather, birds, etc.). Instead of each pulse being processed individually, the pulses are batched typically 3 to 5 at a time with the reflected replies stored in random access memory (RAM) having undergone analogue to digital conversion and quantisation. The returns are examined on a cell by cell basis. The size of cell is selected such that fixed clutter and slow moving returns will be read into each cell from each pulse, but aircraft returns will have moved cells on each pulse. By building up a clutter map and setting thresholds on a fixed clutter basis in each cell, the presence of an aircraft producing a signal above this threshold can be detected.

Other means of reducing precipitation clutter are by means of circular polarisation and choice of frequencies. The latter is usually a compromise since, for ATC, high resolution and high update rates are ideally required. Lower frequencies are less prone to clutter but need correspondingly larger antennae to produce the same resolution as a high frequency radar.

Typically, for approach control, radars of 3 GHz frequency rotating at 15 rpm are usual and for airways surveillance, radars with a frequency in the region of 1.3 GHz rotating at 5-10 rpm are used.

Whilst primary radars capable of measuring height have been in military use for many years, they are not in widespread ATC usage. This in part is due to the availability of SSR (see below), but apart from expense it remains a mystery as to why many more 3-D radars are not in ATC service. It is expected, however, that as technology improves and becomes less expensive, wider deployment will result.

Secondary Surveillance Radar The principle of SSR, otherwise known as Air Traffic Control Radar Beacon System (ATCRBS), differs from primary radar in that instead of depending on reflected pulses, it relies on a transponder in the aircraft which transmits a 12-bit coded response on a higher frequency in reply to the SSR ground station transmission. SSR has provision for up to four sets of data to be coded from an aircraft but in practice only two are used for identity and aircraft altitude. An advantage of SSR is that it is not prone to 'clutter', but a marked disadvantage is that it can only detect the presence of aircraft with working transponders. For this reason, it usually supplements primary radar providing controllers with positive identification and constant altitude information on their radar displays.

Despite its major drawback, there are radar installations with only SSR to provide surveillance of aircraft which are known to exist. For example, as aircraft in airways must pre-notify and receive clearances, and also be fitted with working transponders all aircraft in the airway will be detected. SSR can provide radar surveillance at much lower cost than primary radar due to less expensive transmitter and processing techniques. It is planned that the number of SSR only installations will increase in the future and the number of new primary radars will decrease as a consequence. With the emphasis on the importance of SSR, a number of programmes to provide improvements have been instigated. These encompass improved digital processing and bearing measurement techniques and a development known as Mode-S which includes the ability to selectively address the aircraft instead of on a broadcast basis. The intention of Mode-S is to overcome inadequacies in SSR caused by interference from adjacent aircraft and at the same time provide a data link through increased information in the interrogation and reply and by sending or receiving multiple messages. Mode-S is likely to be introduced in the late 1980s. It is designed to be fully compatible with SSR and thus can be gradually introduced into service. A typical radar configuration with the primary radar and an advanced design of SSR antenna on-mounted is shown in figure 2.

Figure 2. Long range airways surveillance radar (Photograph: Cossor Electronics Ltd.)

Satellite surveillance is a future possibility, but there are no active plans to introduce a system although future air navigation systems are being reviewed by ICAO.

The main use of computers in radar is for processing received signals. The advent of microcomputers and cheap RAM has enabled digital signal processing to advance rapidly improving the quality of radar information. Another significant advance, has been in providing computerised display systems which accept and process digital radar data.

The architecture of radar processing systems varies according to the application. Systems for airways control have to deal with more radar inputs, the need to form composite pictures from those inputs and to process and display large numbers of aircraft at up to 100 control positions. Systems for approach control rarely have more than two radar sources to process and a smaller geographic area is covered with corresponding less aircraft position reports to be contained within the system database.

Design philosophies between manufacturers vary, with a generally accepted view that the display station should have its own processor and store to handle refresh and control functions. The processing of radar inputs and other communications functions can be handled with functionally separate processors or within the display processor in a distributed system which usually offers better safeguards against total system failures.

A combination of analogue and digitised radars can be selected at each display although it is normal practice for an approach controller to select an analogue primary radar input with a digitised SSR overlay (see above) providing data blocks of the aircraft's identity and height. In addition computer generated maps of geographic and aeronautical data can be displayed together with tables of aeronautical data such as the aerodrome weather conditions thus giving the controller a comprehensive view of the air situation. A typical radar display is shown in figure 3. The geographical and airways maps are clearly shown with the aircraft positions and information labels.

AUTOMATION

Automated Emergency Procedures

Emergency procedures are designed to provide the fastest response possible whilst ensuring the continuing safety of the aircraft in the ATC system.

Although most aircraft are well equipped with duplicate R/T sets and antennae, communications failure involving total or partial failures of either the transmitter or receiver, can occur. In the event of communications failure, the air crew can select a reserved

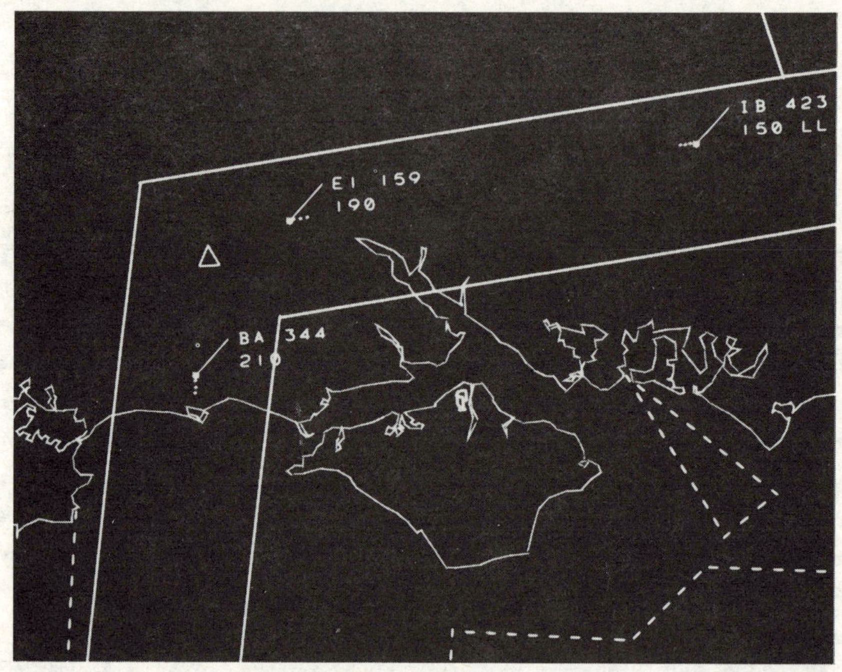

Figure 3. Radar display of airspace over Southern England. The information label shows the aircraft callsign, height in thousands of feet and the destination indicator. (Photograph: Plessey Radar Ltd)

code on the SSR transponder to advise ATC of its predicament. The aircraft should proceed according to its flight plan and ATC controllers will keep other aircraft away from its flight path. Other emergencies in flight can be attended to by Air Traffic Control by expediting the flight in the event of fuel shortages or other unserviceabilities. The aircraft in emergency will also set a specified code on its transponder to indicate that it has a problem. A VHF R/T frequency is reserved exclusively for aircraft emergencies and there are direction finding facilities to determine positional data from transmissions on this frequency.

The systems which relay direction finding information on aircraft in distress are either ground based or satellite based. Ground based systems rely on a network of receivers and direction finders tuned to the emergency frequency. A transmission can be picked up by two or more stations and the position determined by triangulation. To provide full low-level coverage requires a large number of stations and many ATC administrations lack comprehensive coverage of their airspace. As an alternative the satellite system COSPAS/SARSAT has been in experimental operation for a number of years and has proved its worth particularly in remote areas of the world. The satellite detects a transmission on the emergency frequency and relays it to a ground station which can determine the position to within 10 nms. Further development of the system is proposed that will provide the identity of the aircraft making the transmission and some information on the type of distress. It is expected that world-wide coverage could be achieved through installing the locator system on a number of satellites including the INMARSAT series. It is expected that positions can then be determined to better than 2 nm.

Automated Flight Plans

Within an automated ATC system it is clearly important to ensure that the flight plan data is captured and presented to the right controller at the correct time.

With manual ATC systems the flight plan arrives at the Air Traffic Control Centre (ATCC) via the AFTN and information for each sector is manually written or printed on paper tickets, termed flight strips which are then physically passed to the relevant controllers. (See figure 4).

The controller manually updates the flight strip as the flight progresses. Should another controller wish to obtain information on the flight then he will need to communicate by telephone with the responsible controller to obtain up-to-date information.

Within automated systems the flight plans are received via the AFTN with the system checking the message for validity and calculating the times at which the aircraft will reach reporting

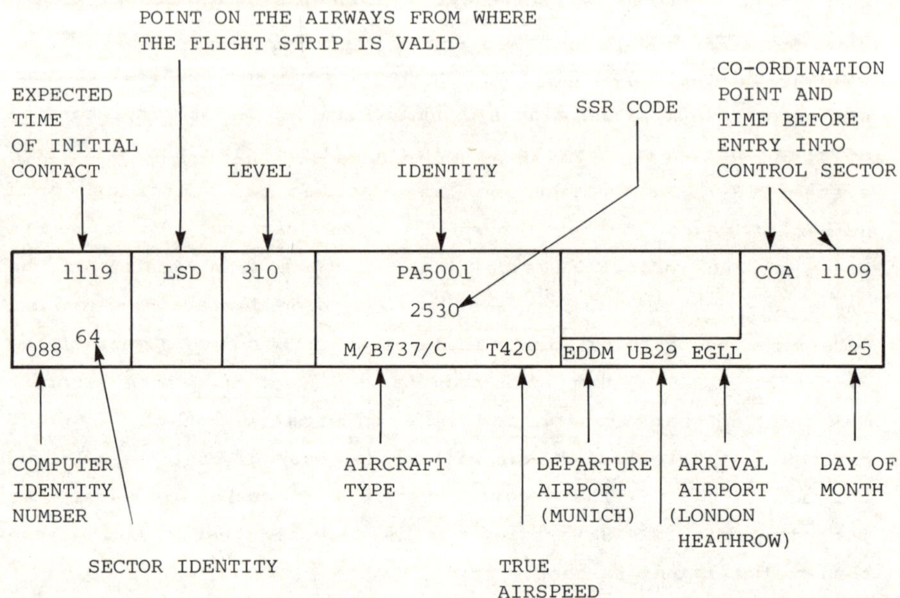

Figure 4 Printed Flight strip from the Clacton Sector at the London Air Traffic Control Centre (supplied by the Civil Aviation Authority).

points having made allowance for wind effect from data entered into the system. Flight plans can also be received by telephone or from aircraft in flight in which case they are manually entered into the system via a terminal. Additionally, since many flights are made at the same time on a regular basis, repetitive flight plans can be stored in the systems database, to be activated at preset times. The flight data is then routed to controllers with the relevant information on the flight through their sector.

In parallel with these developments, flight strips are being superseded by Electronic Data Displays (EDDs) with the advantages that data can be presented in a form which is more legible and easily changed. A significant improvement is that the controller can update any changes by means of a keyboard. A central database is maintained with the current data readily available to any other user in the system. Software Sciences of Farnborough have designed and developed an automated Flight Data Processing System which provides flight data on electronic displays which replace the traditional flight strips. The system is located at the Oceanic Area Control Centre (OACC), Prestwick, and is for the control of aircraft over the busy North Atlantic Ocean. (see figure 5)

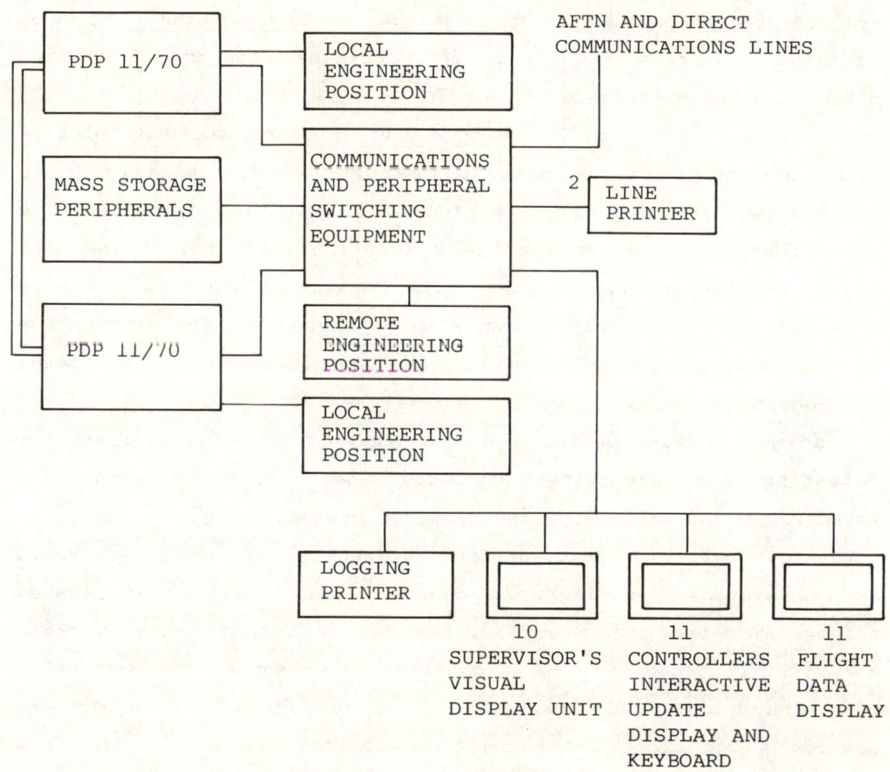

Figure 5. The Oceanic Area Control Centre system configuration.

In the absence of radar surveillance, procedural control techniques are used. Aircraft wishing to use the same route are therefore separated vertically or at time intervals longitudinally. Because of the need to communicate via HF and also to include tolerances for navigational inaccuracies, adjacent routes are separated by 60 nautical miles laterally.

The OACC system accepts direct inputs of weather data twice daily in a form which covers the North Atlantic in a grid. From this, minimum time paths are calculated to define the most economical routings and to enable the system to correct flight plan timings for differences in speed caused by wind.

Flight plan information is received from the AFTN. The messages are accepted automatically, but messages containing transmission or credibility errors are routed to a terminal for manual editing prior to being processed.

Each control position is equipped with an Electronic Data Display (EDD) on which flight details are presented. In addition an Interactive Update Display (IUD) allows the controller to communicate with the system via a function keyboard. Through the IUD the controller can update flight details as they change. More detailed flight information can also be requested. The data on the EDD is capable of displaying data grouped either on a flight level or geographic basis.

A unique feature of the system is that it contains routines for detecting and resolving conflicts where aircraft separation standards would fall below the required minima.

When aircraft are first notified as entering the airspace covered by the system, the flight is checked against all other cleared flights to ensure that its routing does not conflict with other aircraft. In the event that a conflict is detected, the controller can either enter his own resolution or request the system to provide a conflict free path. Once cleared for a conflict free path, should the aircraft's estimated times at position reports change or its route vary, then further conflict detection is performed and the controller alerted for resolution action to be taken should this prove necessary.

The system will also monitor aircraft and controller activity to ensure that time initiated actions take place. Should an aircraft not make a position report at a specified time, the controller is alerted to make contact. If a controller does not respond to system outputs, again alerting messages are made.

In the unlikely event of the system's computers failing, the controller is able to fall back on the data stored at his EDD. The EDD also has battery back-up which can power the display in the event of complete power failure. The controller can enter new flight data manually and keep a record of all flights under his control until the system is restored to full functionality.

Interaction between the ATC system and the human controller

Despite increasing levels of automation in ATC, the controller remains the decision-maker. The role of automation has been to provide the controller with assistance in reducing work loads and enhancing the presentation of information giving greater clarity and ease of comprehension of flight plan and radar data.

Automation has not yet attempted to take over the role of decision-maker although current technology does provide the basis for development. For example, conflict detection systems are now in use to alert controllers if aircraft separation standards are infringed. The Software Sciences OACC system (see above) also offers conflict resolution advice to the controller. It is a comparatively simple step to replace some of the controller decision making in routine situations but because of the very high safety standards required it is unlikely that the controller will be replaced in the provision of an anti-collision service until an entirely fail-safe system can be developed.

Automated systems therefore have to be planned and implemented with the human controller as the key component with the man/machine interface being of prime importance.

The replacement of manual systems often requires a re-appraisal of the true operational needs of the ATC systems as automation of manual practices which have evolved over the period of years is often neither practicable or desirable.

An additional problem with automation is the acceptance of new technology by controllers. An example is the chauvinistic attitude to flight strips which require updating in manuscript. Controllers tend to be sceptical about an Electronic Data Display (EDD) requiring updates through a keyboard, advancing a number of reasons why flight strips are preferable. These include direct updating without the use of keyboard, and a hard-copy fall-back capability in the event of equipment and power failure. The clear advantages of an EDD are that the displayed information forms a part of the system database which can be readily updated and accessed by other controllers. The information is also in a set format avoiding the problems of flight strip mis-interpretation through poor manuscript or idiosyncratic updates. EDDS can be duplicated and provided with battery back-up to provide a fall-back capability but updating using a keyboard or other interactive device remains a contentious point.

(4) The essential requirements of an EDD keyboard are the same as for any other interactive device such as those for control of radar displays, communications devices and displays for weather and supplementary information. The need is to be able to input data or make a selection rapidly and accurately with a minimum number of operations without distracting the controller from his prime task.

Selection devices can either be dedicated to each equipment or as is becoming modern practice, several functions are combined into one selection device. For example a single keyboard can be used for control of radar display facilities in addition to updating a flight data EDD.

As an alternative or as a supplement to alphanumeric keyboards, dedicated function keyboards are often used with the advantage that only single key depressions are needed to select either a communications line or to manipulate data on a radar display. A disadvantage of the function keyboard is the tendency to produce an unwieldy device if a large number of functions are required leading to possible selection errors.

Both alphanumeric and function keyboards can be replaced by a device such as a plasma panel display with a touch sensitive overlay to detect controller inputs. The prime advantage of a plasma panel is its ability to be a programmable keyboard providing any legend or menu sequence desired. Plasma panels are now used for combined radar display and EDD manipulation, data entry and selection of R/T and telephone lines.

The presentation of data to controllers is a subject that has received considerable research effort to determine optimum EDD layouts, character shapes and the use of colour to assist in sorting information and for attention-getting. Monochrome displays are in general use for radar and flight plan displays. The widespread use of colour has been delayed because of the high costs associated with providing colour on radar displays and doubts on cost-benefits in improving controller performance. Tests have shown that whilst controllers have a preference for colour it does not necessarily enhance efficiency and in some instances can be disadvantageous. An illustration is the use of red and green. Traditionally red has been used to denote danger and green for safety. As an attention getter, red is the correct choice psychologically. However the eye is less responsive to red than green and the use of red to alert the controller is therefore not an ideal choice. Nonetheless it is anticipated that colour displays will gradually supersede monochrome displays for all applications, particularly as the general technology trend in computer terminals is towards colour.

Other trends in man-machine interface technology could be the use of voice-recognition systems to change displays, select devices and update databases. The current state of the art in voice recognition will need to advance considerably particularly to avoid mis-interpretation but here the fairly standard and limited vocabulary of ATC will be an advantage.

The development of expert systems heralds interesting (5) possibilities and it is expected that some tasks currently undertaken by controllers could be completely automated. Strategic air traffic management is particularly suited to automation. The concept of air traffic management is to schedule departures (delaying take-off) and arrivals (speed control) to minimise controller work-load and to save fuel by avoiding delaying manoeuvres such as holding in stacks.

Systems currently exist to regulate departures and research into speed control has indicated that expert systems could introduce increased efficiencies producing fuel savings. But as indicated above for safety reasons it is unlikely that expert systems will be used to automate the other control and monitoring tasks for which the human is better equipped.

Simulation

Simulation is used widely in Air Traffic Control based on the straightforward premise that if mistakes are to be made in a learning or training role then they cannot be made with aircraft and people. Simulation is used for training novice controllers, for testing new techniques and equipment, testing of new airspace layouts and procedures to assess the effect on both the control system and air traffic controllers and for refresher training on lesser used techniques such as emergency procedures.

Conventionally, a simulator is based on general purpose computers programmed with the features of the live system. The simulator has to make provision for 'aircraft' to be flown with real flight characteristics and these are flown from VDUs by an operator who can initialise a flight and input changes such as height and course in response to instructions from a controller. Additionally, uncontrolled traffic acting as a background can be provided from a previously constructed database.

The overall control of a simulator is supervised from an instructor position where the type and complexity of the exercise is controlled. The degree to which a simulator has to provide realism varies according to the requirement. For example, a novice controller being introduced to the techniques of manoeuvring aircraft under radar control can practise the basic skills with a simple personal computer with a graphics capability to provide a basic representation of a radar display. There is no need to simulate R/T communications and sitting alongside an instructor who makes the keyboard inputs provides instant feedback and guidance. This type of simulation technique can also be used for on-the-job and refresher training and represents a low cost trainer.

At a more advanced level, simulators provide a control position for the trainee which can be identical with the live system. However, in practice, to achieve a cost-effective simulation, less complex and cheaper display and communications equipment is often substituted without compromising the effectiveness of the training. The simulation of new techniques and procedures is an important role for Air Traffic Control systems which are constantly evolving to cope with the demands placed upon them.

The implementation of changes in airspace layouts can be modelled on a computer to give an indication of the effects of introducing a change. The effect of changes on controllers, particularly with regard to work load and the co-ordination with controllers of adjacent sectors, can only be assessed in detail by simulation using equipment similar to the live system and running an exercise as realistically as possible. The exercise can be analysed and, if necessary, a number of candidate proposals evaluated in order to provide a basis for selecting a new procedure for adoption. In a similar vein, new equipment can be tested to assess the acceptability and ergonomics before deployment in live systems.

In preference to purpose-built simulators, ATC systems often have a capability to provide the basis of a simulator. Provided that there is sufficient hardware capacity, part of the system can be allocated to a training role, either on a permanent basis or as required. Where training demands are not onerous, this can be a very cost effective technique in respect of not requiring additional maintenance and software support.

FUTURE DEVELOPMENT

Predicting trends in Air Traffic Control particularly with respect to likely timescales is not easy. The need to agree new standards on an international basis, the long planning cycles and high cost of major systems programmes tends to exclude revolutionary concepts. The systems approach to date, has been one of evolution, improving concepts and equipments within an established framework.

Equipment developments will continue with improvements in performance and reliability.

Fault tolerant computers are already in operational ATC service and it is expected that this type of architecture will gradually supersede the traditional main and standby approach.

Improvements in high resolution computer graphics terminals will provide a challenge to the specialist and expensive radar displays with colour will be in more widespread use.

Communications, navigation and surveillance equipment will rely less on custom-designed hardware and increasingly on standard processor modules with specialist firmware.

New surveillance systems such as Mode-S will gradually be introduced from the late 1980's. A new microwave landing system will be introduced in the late 1990's. Satellite based systems for communications, navigation and surveillance will not be in general use until after 2000 although some initial usage will occur from 1988 onwards.

The next major systems advance will be in developing the concept of strategic Air Traffic Management. This will be coupled with higher levels of automation with the intention of providing a high level of efficiency and fuel conservation as air traffic increases whilst at the same time providing the high level of safety which the aviation industry demands.

EDITORS' NOTES

(1) R/T refers to VHF radio telephony, which is the universal communications medium between ATC centres and air crews.

(2) Packet switched networks with X25 protocol are now a world wide standard.

(3) The use of satellites for navigating land-based vehicles is discussed in Chapter 14.

(4) Chapters 10 and 12 allude to the problem of user reluctance to adopt the new technology in the bus and travel industries respectively.

(5) See Chapter 15.

Information Technology Applications in Transport, pp. 191–210
P. Bonsall, M. Bell (Editors)

Chapter 9

SOFTWARE FOR ROAD FREIGHT ROUTEING AND SCHEDULING

Michael Sahling
PTV GmbH, Federal Republic of Germany

INTRODUCTION

Road freight traffic can be broadly classified in two ways

a) long and short distance road freight
b) own account operation and public haulage

For long distance freight, whether carried by own account operator or by public haulier, only a limited number of destinations is generally to be visited. In contrast, for short distance freight the problem is usually to service a large number of customers lying close to each other within a limited time interval. The resulting vehicle routing and scheduling tasks are therefore of varying types and require, both at the level of within and between business communication as well as at the level of routing and scheduling, different kinds of solution.

Since the routing and scheduling problems for short distance road freight are less complex than those for long distance road freight, the development of computer based planning methods has been concentrated in the past essentially on the former area. However, it is here that the more complex public haulier routing and scheduling problems are to be encountered, because the vehicles are used to collect and deliver freight simultaneously. In the case of own account operations, vehicles are generally used for the distribution of freight only (see Fig. 1).

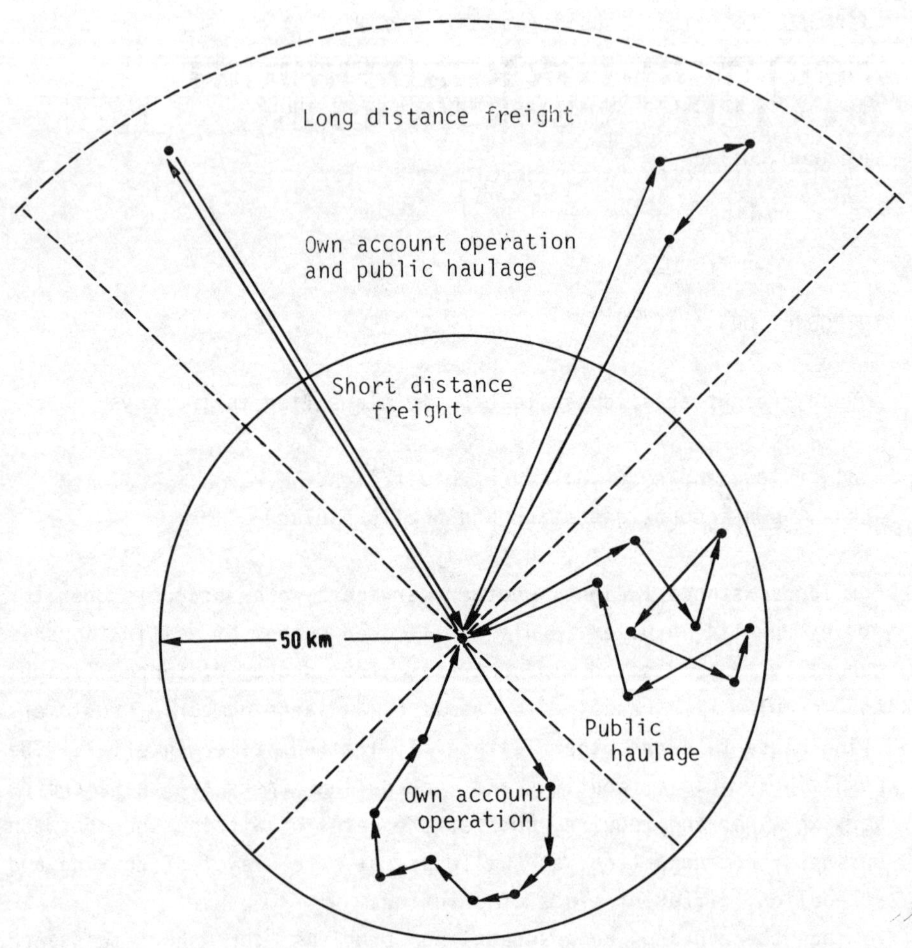

Figure 1: Structure of the freight distribution problem

The simplist type of routing and scheduling problem referred to above, for example the delivery of beer from a brewery to its customers, will be considered in the following in more detail since it is here that computer based planning methods are most developed. After a concise review of the underlying algorithms, the computer programs based on these will be briefly described. Finally anticipated future developments for more complex routing and scheduling problems will be discussed.

PROBLEM DEFINITION

Planning the operation of vehicles in a fleet requires the solution of difficult combinatorial optimisation problems. These are known in general as routing problems.

Routing problems may be divided according to possible solution procedures into two groups;

- link oriented problems; and
- node oriented problems.

The first class of problems has the following form; a tour starting from and ending at a particular point, and making use of a number of given links, must be determined so that, subject to certain conditions, its length is minimised.

The following practical problems are of this type;

- the postman´s problem;
- the refuse collection problem;
- the street cleaning problem; and
- the general street maintenance and patrol problem.

Problem formulation and solution methods for this class of problem will not be considered here, since for the distribution of goods they are of no significance.

In contrast, the node oriented problems have the following form;

a tour starting from and ending at a particular point, and including a number of given nodes, is determined so that, subject to certain restrictions, its length is minimal.

Suitable solution methods for this class of problem will first be examined in more detail from a theoretical standpoint.

Methods for computer based node oriented route planning have their origins in the so called travelling salesman problem. One of the first tours to be planned by mathematical optimisation methods was an election campaign trip of an American presidential candidate who in the 1950s wished to visit a particular selection of large American cities in as short a time as possible. Here we have a simple sequencing problem, whereby the sequence in which the cities are visited must be chosen so as to minimise total flying time.

With a small number of locations or nodes this and similar problems may be solved exactly. For a large number of locations the optimum can, only in exceptional cases, be found because there are to date no exact optimisation techniques that can solve problems of such a degree of combinatorial complexity within a reasonable amount of computing time. This is particularly the case when not only an optimum tour but also the operational schedule for a large fleet of vehicles is to be determined. This is a combined assignment and sequencing problem whereby on the one hand contracts are to be assigned to vehicles and vehicles to tours, while on the other the sequence of stops at customers within a tour is to be determined. The potential for optimisation is, in general, largest when these tasks are solved simultaneously.

In the case of problems of practical interest there are additionally a large number of restrictions to be taken into account. The most important are the following:

- vehicle capacities;
- working or operating times for drivers or vehicles;
- different loading devices; and
- customers´ hours of business. (1)

Since routing and scheduling problems for a large vehicle fleet, where there are a wide variety of restrictions, are not solvable within a reasonable amount of computing time by an exact optimisation procedure, heuristic methods have in the past been developed which, taking into account different rules, yield good approximate solutions.

The state of the art and current research with regard to theory and methodology is documented in a large number of mostly English language journal articles. The most extensive review to date of the various types of problem and their solution procedures is given by Bodin et al. (1982), which also includes around 600 references. For further information on methodology, the interested reader is referred to it.

In the past, four classes of procedure have proved to be useful in practice

- one-stage methods
- two-stage methods
- improvement and exchange methods
- exact methods.

One-stage methods

The savings method of Clarke and Wright (1964) is the most well known one-stage solution procedure and forms the basis of many commercial programs. The savings method combines customer locations by the calculation of savings and solves the problems of the assignment of clients to tours and the sequencing of clients within tours simultaneously.

Initially it is assumed by the savings method that each node or client is served from depot 0 individually. Thus each tour or round trip contains one node only, and the number of tours is equal to the number of nodes. For this worst conceivable initial solution we have, assuming symmetry of distance ($d_{ij} = d_{ji}$), the following total distance:

$$D = 2 \sum_{i=1}^{n} d_{oi}$$

By the successive aggregation of tours, the algorithm seeks to improve the solution (reduce D). Assuming triangular inequality, the combination of two tours must achieve an improvement (see Figs. 2a and 2b). The combination of tours according to the savings method must occur so that the start and end nodes of two different tours are connected to each other and the connections to the depot are eliminated.

Through this connection a saving of

$$S_{ij} = d_{oi} + d_{oj} - d_{ij}$$

is achieved, as shown in Figs. 2a and 2b.

The savings indeces for all node pairs are calculated in advance and ranked according to magnitude. Tours are so constructed that at each step of the procedure, the permissable connection (v_i, v_j) is chosen so that when the corresponding connections to the depot are out the savings index is maximised. In this context a connection is permissable if the following conditions are fulfilled;

- v_i and v_j belong to different tours;
- v_i and v_j are the start and end nodes of tours;
- the capacity restriction of the vehicles and the time restrictions for the tour are not infringed; and
- the savings index s_{ij} has positive value.

The method terminates when there are no further permissable connections.

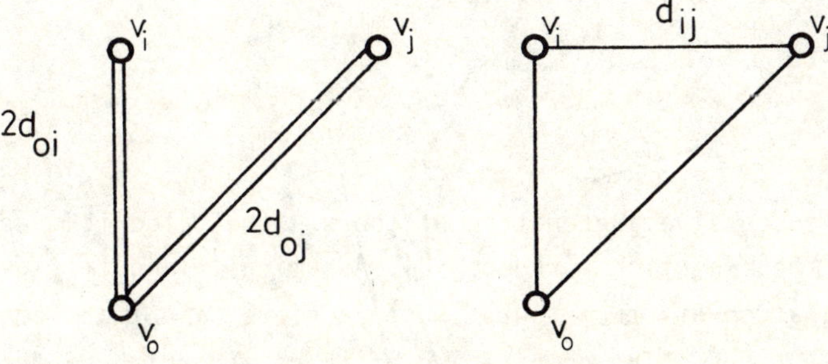

a) Initial solution and first connection

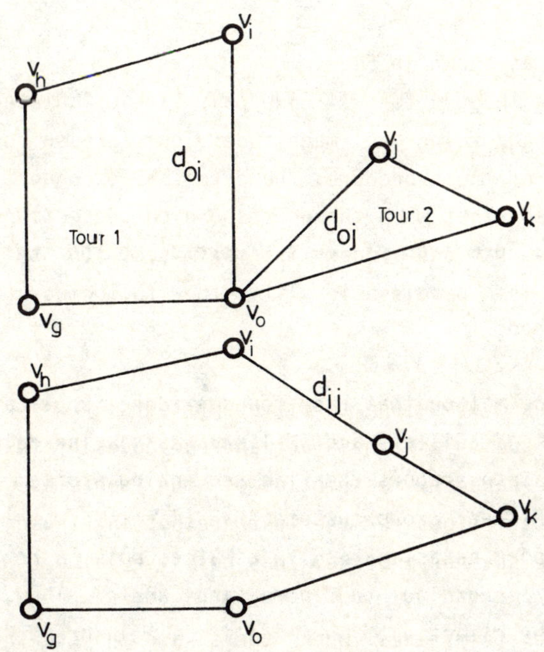

b) Conditions before and after the connection of two tours

Figure 2: Formation of tours according to the savings method

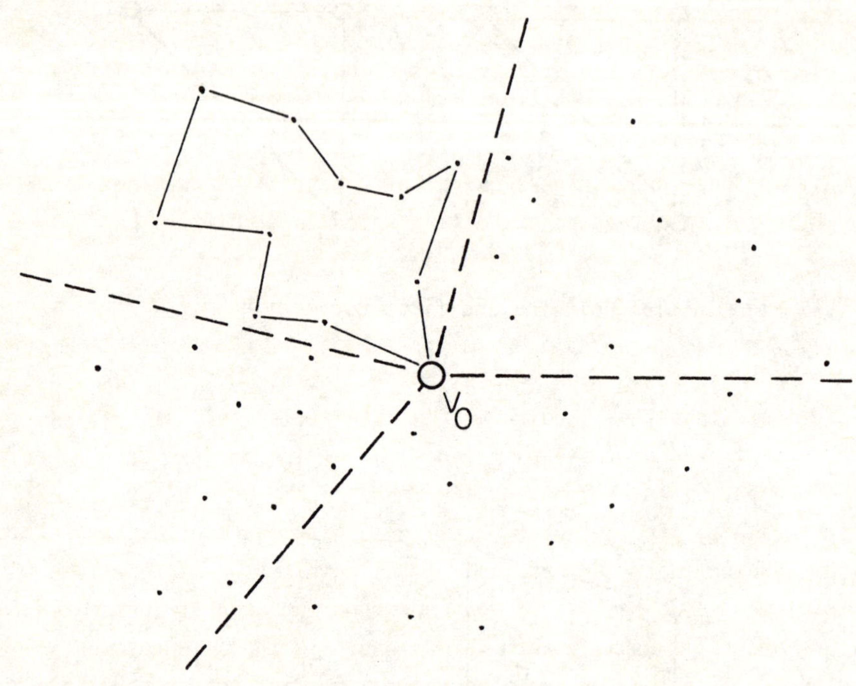

Figure 3: Formation of tours according to the sweep method

Two-stage methods

The two-stage method that has found widest application is the sweep method of Gillet and Miller (1974). The delivery area is first divided into sectors then the sequencing problem is solved. Gillet and Miller group customers v_i ($i = 1, \ldots, n$) according to their polar coordinates, taking V_0 as the pole. The clients are first sorted according to increasing angle. Thus, if α_i is the angle formed by client v_i, then $i > j$ if $\alpha_i > \alpha_j$ or if $\alpha_i = \alpha_j$ and $d_{0i} < d_{0j}$ ($i, j = 1, \ldots, n$; if $i \neq j$). Adjacent clients in the series are then grouped, starting with the first, until the inclusion of further customers infringes capacity restrictions. Thus, different groupings may be produced by turning the axes so that different customers assume first position in the list of clients.

Figure 3 shows an example. The clients lying within the sectors delineated by the dashed lines form groups.

Within these groups the sequence of clients is determined as the solution of the standard travelling salesman problem.

Incremental improvement methods

These methods seek a stepwise inprovement of an initial solution. Since they require an initial solution, as yielded by the savings or sweep methods, they cannot be regarded as independent planning methods. An improvement occurs by changing, for example, the sequence of clients within a tour, by exchanging individual clients or chains of clients between tours, or by dissolving and reconstruction neighbouring tours. The search for improvements is usually terminated when further improvements are nolonger expected or when a certain computer time threshold has been exceeded (see, amongst others, Christofides and Eilon, 1969, and Lin and Kernighan, 1973):

Exact methods

As already mentioned, exact optimisation procedures applicable to practical problems have yet to be developed. The largest problem to be handled by an exact optimisation procedure so far reported in the literature contains 31 clients (see Christofides, 1976). Christofides, a defender of exact solution procedures, has developed a special branch-and-bound algorithm for larger problems, which nonetheless has to be prematurely stopped to avoid unreasonably large computing times (Christofides et al., 1979). One should therefore, at least at present, talk only of incomplete exact methods. It is to be anticipated, however, that with improving hardware performance exact solution procedures will become increasingly relevant for applications.

PLANNING DATA AND RESTRICTIONS

The optimisation algorithms, as outlined in the preceding sections, must be fitted into a programming environment so that a user friendly planning tool is created. For this reason, all commercially successful program packages are similarly constructed (see Fig. 4), so that only some essential components of the planning data and restrictions will be discussed here. These are:

- the planning area;
- the customer data;
- the contract data; and
- the available vehicle fleet.

The planning area

The planning area encompasses the geographical space in which all customers lie. This area can be specified in two ways:

(i) The coordinate method
The locations of customers are described by their coordinates. The distances between customers are calculated on the basis of airline distances. Actual (road) distances and journey times are arrived at by means of detour factors and mean speeds.

(ii) The network method
The locations of customers are described in relation to a road network. For each link in the network, length and mean journey time is specified. Customers are located at nodes or on links. The distances and journey times between customers are calculated by a shortest route algorithm.

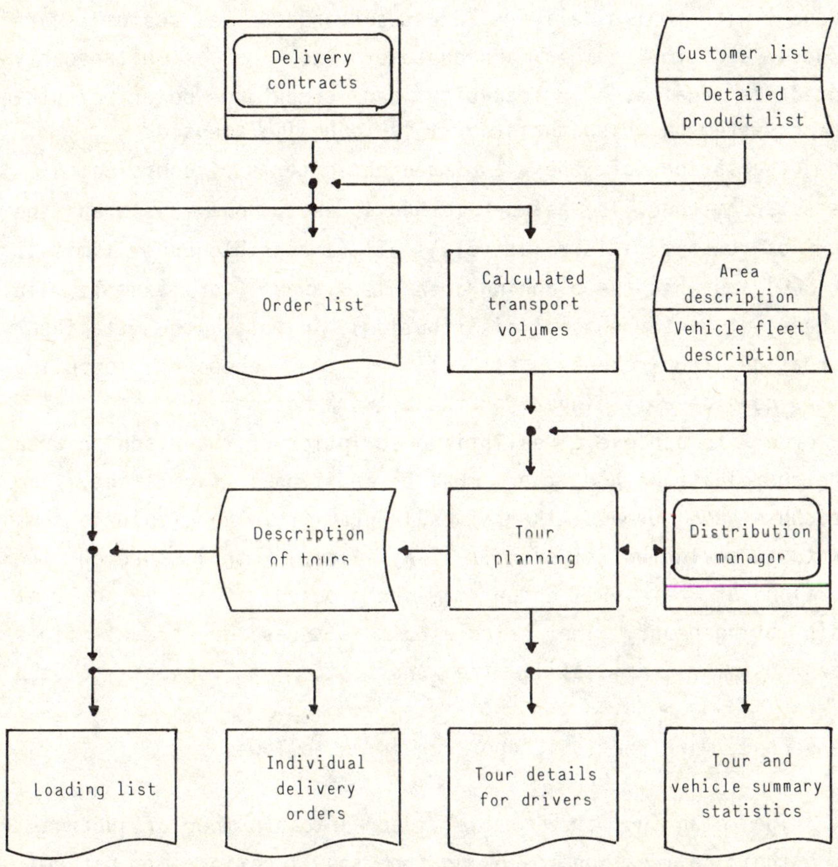

Figure 4: The role of tour planning within the freight distribution process

The choice of suitable method is based on a number of criteria. Particularly important are the size of the planning area, the frequency with which delivery adresses change or new customers are included, and the number of customers to be simultaneously included as against the capacity and computing power of the computer system on which the program is to be implemented.

The description of the planning area in terms of coordinates is the standard method. The network method is used, however, when the number of customers is relatively small (a sensible upper limit is about 500) and when the planning area is large (for example, in the case of interregional distribution). In total, the data input and updating requirements, as well as the necessary computing time, is considerably higher for the network method.

In order to achieve a realistic description of the planning area by the coordinate method, there must be additional facilities. Two approaches have proved themselves in practice. The simpler allows for natural obstacles (rivers, railway lines, etc.) through the imposition of barriers which may be crossed, if at all, only at certain bridgeheads. The calculated distances or travel times between customers are still the shortest, but subject to the barriers.

The second, more realistic approach is as follows:

(i) The planning area is divided into a number of subareas, their delineation following from the location of natural, constructed or other forms of barrier to traffic. Separate detour and mean speed factors may be specified for each subarea.

(ii) Restrictions on border crossings may be imposed on tours in order to take into account the effects of natural or other traffic barriers. Thus it is possible to restrict the connections between certain groups of subareas to specific locations.

(iii) For direct connections between subareas, alternative detour and mean speed factors may be defined in the form of a subarea matrix.

Customer data

The customer data describes the location of all customers within the planning area as well as particular customer requirements, for example delivery or pickup times and vehicle size. This data is produced just once, and only with the inclusion of new customers or with the updating of customer requirements is it changed.

Two methods for describing the locations of customers have proved useful in practice:

(i) For large area planning, coordinate locations may be determinded with the help of various data banks from postal codes and town or district names.
(ii) In the case of urban areas, customer locations can be entered directly on a map. Coordinates can then be read off manually or with the assistance of a digitiser.

Contract data

Tours planned depend on the contract data. For all the customers included, the collection and delivery quantities, as well as particular restrictions (concerning, for example, delivery times), have to be taken into account. All customers included in the contract data must have been defined in the customer data. However, only a small proportion of the customers entered in the customer data are in general involved in contracts which must at any one time be executed.

Vehicle fleet description

The available vehicle fleet often consists of different vehicle types. Each vehicle type is generally identified by the following factors; the vehicle size, operating time and type specification. Vehicle size may be described simultaneously by more than one kind of unit (for example, weight, number of paletts, or volume). The operating time defines the period of availability of the vehicle. All tours must be completed within this period. The type specification permits a classification of the vehicles in terms of customer restrictions or special features of the load. Thus the required vehicle occupancy may be given, as well as the availability of a refrigeration unit.

PROGRAM PACKAGES

Since the early seventies, numerous commercial program packages, based on the operations research methods, the planning data and the restrictions described earlier, have been developed. However, only a limited number have received wide application.

The detergent and food manufacturer Unilever Ltd., London, was one of the first developers and users of freight distribution software in Europe. Corresponding to its early stage of development, the program ROUTEMASTER, which uses the savings method, was designed for batch operation as a strategic planning tool, primarily for the Unilever subsidiary Walls Icecream and Birds Eye. This program must now be regarded as obsolescent since it cannot be used for daily operational planning in an interactive mode.

One of the first developers of freight distribution software on the continent of Europe was the firm of Siemens with their program package TRAFIC. Here also strategic planning in batch operation was the primary objective. However, after extensive changes between 1983 and 1985, the program was modified for daily operational planning. It should, however, be mentioned that TRAFIC is only offered for Siemens hardware.

Program packages developed in the 1980s have the advantage that the significant benefits brought by increasing computer power and interactive operation could be fully exploited. Nonetheless, to (2) date only a limited number of program packages have achieved widespread use in national markets or at the European level. Table 1 shows the programs that, to the author´s best knowledge, have many successful implementations.

Only those program packages that contain heuristic or exact optimisation methods are included, not however those that merely assign contracts to tours or such like. Also the list would be longer if the program packages designed for strategic use, rather than for daily operational planning where swiftness and interactive operation are required, had also been included. The program MOVER in Table 1 achieves the required degree of swiftness by interrupting the computationally intensive, exact optimisation procedure before completion.

The savings method clearly dominates in practice, both because of its swiftness and its ability to take into account a host of restrictions. For historical reasons, FORTRAN dominates. Personal computer (PC) versions of the market leading programs are either already available or in preparation.

The area of application of all the listed programs is primarily in own account freight distribution, usually relating to one depot and one planning period. Simultaneous multidepot, multiperiod solutions are to date only available in theory, even if the converse has been suggested by program salesmen. The same applies for the application of the programs to the public haulage sector. Individual applications of, for example, MOVER and INTERTOUR to the classical problem of the simultaneous collection and delivery of freight should not conceal the fact that here there is still considerable scope for further development.

Table 1. Freight distribution programs

Firm	Program name	Basis method	Computer language
Battelle-Institut e. V. Frankfurt, Germany	MULTITOUR	SWEEP	FORTRAN
Istel Ltd. Oxford, England	VISIT	SAVINGS	FORTRAN
ORES B. V. Amsterdam, Netherlands	MOVER	incomplete branch-and-bound	FORTRAN
PTV GmbH Karlsruhe, Germany	INTERTOUR	SAVINGS	PASCAL
Siemens AG München, Germany	TRAFIC	SAVINGS	FORTRAN

SUMMARY AND FUTURE PERSPECTIVES

The development of the underlying theory, as well as problem oriented, heuristic algorithms for the computer based optimisation of own account freight distribution, had in the 1970s reached the stage where they could be used by larger firms. The improvements in hardware performance and the largescale introduction of personal computers in the 1980s have put the software and hardware at the disposal of the despatcher at his place of work. A number of comprehensive program packages are being successfully used for the daily operational planning of freight distribution for different firms from various sectors of industry. The integration of such software into complete fleet management software is currently the subject of intensive research (see, for example, Commercial Motor, 1985, Vol. 11, p. 95). (3)

The current situation in the public haulage sector is quite different. Computer aided planning methods are either not used at all or only to a limited extent. The despatcher makes use of the telephone and usually some home grown planning procedure. By these means contracts are made, contracts are assigned to vehicles, and tours are determined.

A further, rarely used, aid is the street map on which delivery areas may be indicated and customer locations marked using, for example, coloured magnets. This describes the state of the art, particularly for middle size firms.

There are four reasons for this lower level of computerisation in the public haulage sector:

(i) In own account freight distribution, the problem is in general to distribute freight. In contrast, in public haulage, the task is usually to collect and deliver simultaneously. For this problem, other algorithms which can incorporate a multitude of restrictions are required. There is here scope for much development.

(ii) In own account freight distribution, optimisation is based almost exclusively on distance or travel time. However, in public haulage, at least in some European countries, the different tarif structures have to be taken into account.

(iii) In own account freight distribution there is in general a fixed set of customers, whereas in public haulage only about 50 percent of customers are regular. Here computer based tools for the generation of planning data must either be created or further developed. For example, coordinate or street data banks have to be extended so that customers may be automatically assigned coordinates or network nodes.

(iv) Computer aided freight distribution planning must, in the public haulage sector, be set within a more extensive information and communications system in order to exploit their full potential. This applies to both within and between firm systems, ranging from simple backload communication and information systems (for example, for removal firms) to more complex central freight bartering schemes (for example, Datafreight in the U. K.).

When these four gaps in development have been closed, which is to be anticipated as a result of intensive research in a number of places, the use of computer aided freight distribution planning will spread in the public haulage sector as it has done in the own account sector.

REFERENCES

BODIN, L. D. GOLDEN, B. L. ASSAD, A. A. BALL, M. O. (1982). Routing and scheduling of vehicles and crews - The state of the art. Computers and Operations Research 10

CHRISTOFIDES, N. (1976). The Vehicle Routing Problem. Revue Francaise d´Automatique, Informatique et de Recherche Opérationelle 10, pp. 55 - 70

CHRISTOFIDES, N. EILON, S. (1972). Algorithms for the Vehicle-Dispatching Problem. Operations Research Quarterly 23, pp. 511 - 518

CHRISTOFIDES, N. MINGOZZI, A. TOTH, P. (1979). The Vehicle Routing Problem. In: CHRISTOFIDES, N. et al. Combinatorial Optimization. Wiley and Sons, Chichester-New York-Brisbane-Toronto, pp. 315 - 338

CLARKE, G. WRIGHT, J. W. (1964). Scheduling of Vehicles from a Central Depot to a Number of Delivery Points. Operations Research 12, pp. 568 - 581

Commercial Motor (1985). On the road ... Vol. 11, p. 95

GILLETT, B. E. MILLER, L. R. (1974). A Heuristic Algorithm for the Vehicle-Dispatch Problem. Operations Research 22, pp. 340 - 349

LIN, S., KERNIGHAN, B. W. (1973). An Effective Heuristic Algorithm for the Traveling-Salesman Problem. Operations Research 21, pp. 498 - 516

EDITORS' NOTES

(1) There are obvious similarities between the problem set out here and the public transport routeing and scheduling problem described in Chapter 10. However, freight scheduling must take account of the greater heterogeneity of the load which necessitates the use of different kinds of vehicle. The problem is further compounded because the different kinds of vehicle may be subject to differing loading and unloading restrictions.

It is also interesting to compare the routeing and scheduling problem faced by road freight operators with that described for rail traffic in Chapter 7. Although the goods carried by rail are more homogeneous and the routeing more constrained, the matching of motive power units to wagons is a significant issue.

(2) An indication of the contribution that may be made by interactive graphics is given in Chapter 11.

(3) Free-standing software already exists for the following purposes: load planning; order capture; depot location; operations costing; vehicle maintenance and replacement; fuel monitoring; tacho analysis; stock control and warehouse operations. Further details of some of these are available in the proceedings of the Seminar on Information Systems in Commercial Vehicle Operations held at Cranfield Institute of Technology in June 1985.

Information Technology Applications in Transport, pp. 211–234
P. Bonsall, M. Bell (Editors)

Chapter 10

SOFTWARE FOR BUS OPERATIONS PLANNING

Anthony Wren
University of Leeds, UK

INTRODUCTION

Computers have been used by bus operating authorities for general management purposes for at least twenty years, but it is only relatively recently that they have been widely applied in the design and operation of bus systems. Potential users of specialist software should beware of adopting a system only because it is user-friendly or produces fine output. They must be satisfied that the results it produces are realistic and efficient. Unfortunately, however, the precise algorithms used in commercial systems are rarely made known, and thus critical assessment is not always possible.

The author has attempted to augment suppliers' descriptions from his own knowledge, and from any published papers relating to the methods used. Given limited space, this has inevitably meant that systems familiar to the author are treated in greater depth.

NETWORK PLANNING

It is clearly important that the manager should have access to advanced information systems to help him develop any new or revised network and assess the consequences of proposed actions. An early network planning project was in Wallasey, UK (see Lampkin and Saalmans 1967), where routes (lines of service) and frequencies were developed by computer from passenger travel statistics.

Several analysts felt that the computer should more appropriately assess candidate networks proposed by management or consultants, and perhaps develop levels of service for these. Such an approach was adopted in Coventry, UK, by the Local Government Operational Research Unit, (see Orford 1974). In this case, a full modal split model was used, including a route choice element. Predicted passengers were assigned to routes of prespecified trial networks. Because modal split and assignment both depend on the level of service, a basic level was specified for each route, using a proportion of the buses available. Further buses were automatically added to those routes with heaviest loadings, and the modal split and assignment stages were repeated, until a target number of buses was reached. The computer then indicated bottlenecks in the system.

The most widely used computer package for network planning internationally is probably the Volvo Interactive Planning System (VIPS), originating in Sweden, but marketed worldwide. This follows similar, but more sophisticated, processes to those used in Coventry, and is aided by good interactive graphical facilities. The network is represented to the computer in terms of routes, frequencies, running times, etc., and various performance indicators are produced, relating to loadings on routes, passenger totals, revenue statistics, and measures of quality of service. After collection and validation of data, VIPS assigns individual passenger trips to paths through the network, passengers between any origin and destination pair being distributed among the paths according to relative attractiveness.

Several networks are evaluated. While these are normally produced by the planner, it is possible to ask the computer to construct a trial network, by a process similar to that used in Wallasey; this may then be modified interactively. The planner may specify frequency of service on the various routes, or instruct the computer to propose frequencies. When several routes pass through or terminate at a particular point, the computer can be asked to show whether passenger travel times or operational efficiency could be improved by breaking and re-linking the routes there.

Implementations of VIPS include various Scandinavian cities, Singapore, and several British locations, including areas of London. Although VIPS includes attractive interactive facilities, it is believed tht these have often not been used by customers on their own premises, the results having been obtained by batch processes. A fuller descripotion of VIPS is given by MVA (1985).

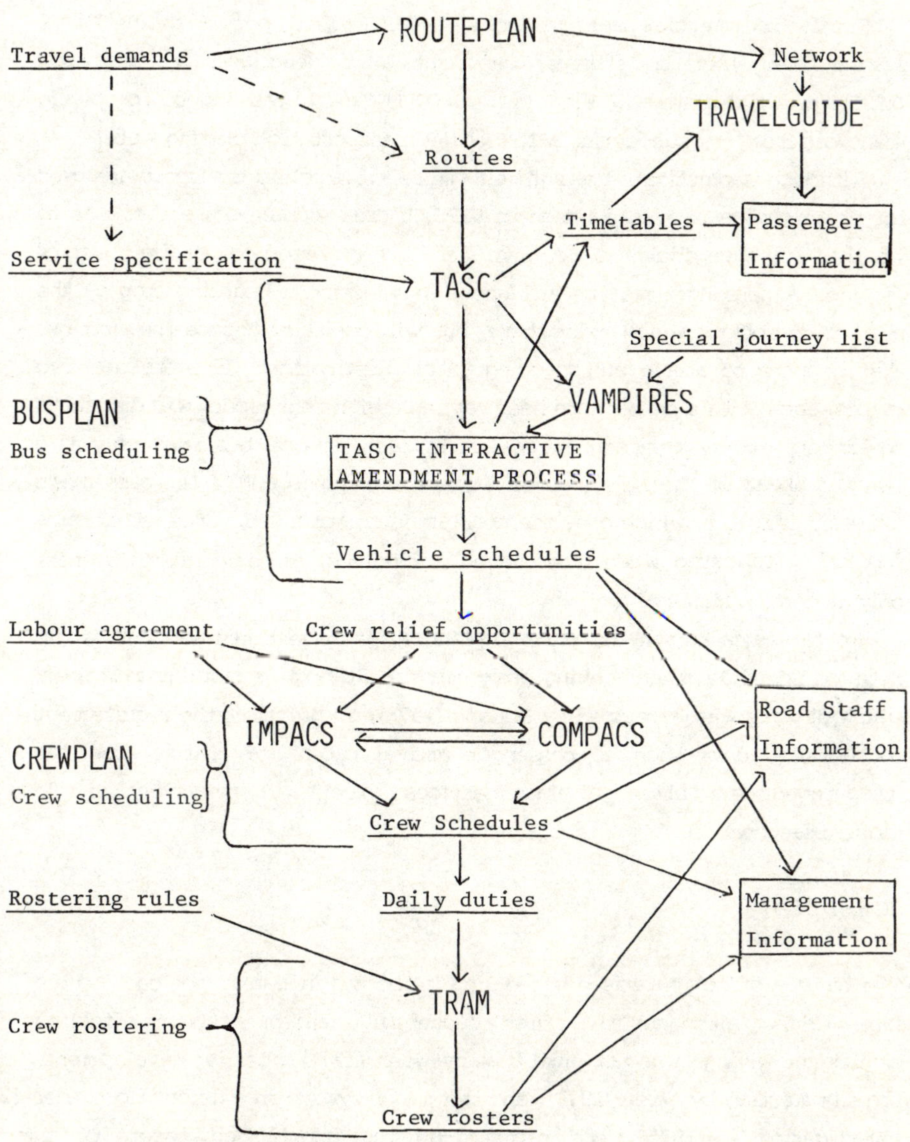

Figure 1. BUSMAN

Constituent processors of BUSMAN are in large block letters.

Similar commercial packages include ROUTEPLAN, part of the BUSMAN system from Wootton Jeffreys Consultants Ltd., described by Williamson (1985), and illustrated in Figure 1. ROUTEPLAN was adaptd for public transport from the successful MINITRAMP transportation planing suite.

Although sophisticated assignment models for public transport are used for route choice in systems such as VIPS, there must be some doubt as to whether any such choice is meaningful if it is not based on precise times of service. A passenger may be influenced in selecting a boarding stop by the number of options which exist there, but will normally choose the first bus that happens to arrive and is going to his destination. If there are two routes, each with a ten minute headway, an assignment model will distribute passengers evenly between them, but in fact, if one bus runs only two minutes ahead of the other, it will attract eighty percent of the passengers (unless it is full, in which case the problem is compounded). The situation is further complicated when interchange is involved, or some routes operate only at certain times of day.

As far as is known, no current planning package takes account of the detailed schedule when making assignments, although a simulation system which did this was developed by Bolland (1978). In practice, the requirement for detailed data such as passenger arrival rates and traffic flow has discouraged simulation of bus services except in relatively simple circumstances.

SCHEDULING

The history of computerised scheduling in the bus industry goes back to the late 1950s, although the first successful implementation effecting significant savings was not until that by Wren (1972). Earlier developments are summarised by Wren (1971), and the major system in existence or under investigation in 1975 were reported in the unpublished papers of an international conference in Chicago. The state of the art up to about 1983 is found in Wren (1981) and Rousseau (1985).

We discuss here the stages of scheduling and their solution by some of the most widely used programs. The main features of these and some other programs are summarised in Figure 2.

	B	G	T	J	C	E	R	P	H	I	M	D	W	V	Z
BUSMAN	Y	Y	Y	Y	Y	Y	Y	-	Y	Y	Y	Y	M	Y	Y
TASC	R	-	Y	-	-	Y	Y	-	Y	Y	-	Y	M	Y	Y
VAMPIRES	N	-	Y	Y	-	-	-	-	Y	P	-	-	M	Y	-
COMPACS	-	-	-	-	Y	Y	-	-	Y	Y	-	Y	M	Y	-
IMPACS	-	-	-	-	Y	Y	-	-	-	P	Y	-	3	-	-
HASTUS	R	-	Y	-	Y	Y	-	-	Y	Y	Y	Y	M	?	?
BUS	R	-	Y	-	-	-	-	-	Y	Y	-	Y	?	?	-
MACRO	-	-	-	-	-	Y	-	-	-	-	Y	-	M	?	-
MICRO	-	-	-	-	Y	Y	-	-	-	?	Y	?	M	-	-
SAGE	R	-	Y	-	Y	-	-	Y	Y	Y	Y	Y	M	Y	Y
MINI-SCHEDULER	R	-	Y	-	-	-	-	-	Y	Y	-	Y	M	Y	Y
RUNCUTTER	-	-	-	-	Y	-	-	-	Y	Y	-	Y	M	Y	Y
RAMCUTTER	-	-	-	-	Y	-	-	-	Y	Y	-	?	M	-	-
RUCUS II	Y	-	Y	-	Y	-	-	-	Y	Y	-	Y	?	-	-
BUS	Y	-	Y	-	-	-	-	-	Y	Y	?	Y	?	-	-
CREW	-	-	-	-	Y	-	-	-	Y	Y	-	?	?	-	-
Belletti et al. (1985)	-	-	-	-	-	-	Y	-	-	Y	Y	?	1	-	Y
Brandani et al. (1981)	R	-	Y	-	-	-	-	-	Y	Y	-	Y	1	-	-
Ceder & Stern (1985)	N	-	Y	Y	-	-	-	-	Y	Y	-	-	-	Y	-
Hoffstadt (1981)	N	-	Y	-	Y	-	Y	-	-	Y	Y	Y	1	-	?
Howard & Moser (1985)	-	-	-	-	Y	-	-	-	Y	Y	-	-	-	-	-
Leprince & Mertens (1985)	-	-	-	-	Y	-	Y	-	Y	?	-	Y	1	-	Y
Li Yihau (1985)	Y	-	Y	-	?	-	-	-	Y	Y	-	?	1	Y	?
Mitra & Darby-Dowman (1985)	-	-	-	-	Y	-	-	-	-	-	Y	?	1	-	?
Piccioni et al. (1981)	-	-	-	-	Y	-	-	-	-	-	Y	-	?	-	-

Key to columns (Y = feature present; dash = feature absent)

B Bus scheduling (N = network-based; R = route-based)
G Timetables generated from demand
T Timetables based on regular headways and direct trip input
J Critical journeys highlighted for possible revision
C Crew or driver scheduling (run-cutting)
E Estimator of crew numbers
R Roster production - rotating
P Picking process management ad weekly work compilation
H Method - Heuristic
I Interactive features (P = partial)
M Mathematical (usually ILP with heuristic reduction)
D Range of working documents produced
W Number of known working implementations 1985 (M = five or more)
V Version available on desk-top micro-computer
Z Database management facilities

Figure 2. Some current implemented computer scheduling systems showing

(1) Timetabling

There are two radically different approaches to the determination of times of buses running on given routes. One builds a timetable according to observations of demand which vary continuously throughout the day. The other uses fixed service intervals which change a few times daily.

Timetabling from demand. Where regular observations are taken of passenger flow past a critical point of a bus route, it is possible to determine service intervals which will in theory result in every bus having an equal number of passengers. A target number may be chosen as a given percentage of capacity, to allow for random fluctuations, and it may be appropriate to use different targets at peak and off-peak times. The result is a set of departure times giving service intervals which change gradually according to demand, although a maximum interval may be imposed.

Timetabling was based on this concept in North America well before the advent of computers, and it was natural to automate the approach. In principle, one can determine suitable bus times in each direction of travel, then match arrivals and departures at terminal points. Indeed, this was done in many programs, for example, by Elias and Smith (1970). However, these early programs did not coordinate times between the two directions, and could therefore use too many buses. The author has not been able to determine whether current American systems operate differently, but unpublished student work has recently led to programs overcoming the problem.

In practice, the situation is complicated if routes bifurcate, or some trips work only part of the route, and it is necessary to ensure proper integration of the various service patterns. For this reason, programs such as the MINI-SCHEDULER from SAGE (1985) have been developed to build up vehicle workings interactively.

Timetabling regular headways. In many countries, buses operate at fixed intervals and constant scheduled running times for considerable periods of the day. At the crest of the peak, additional journeys may be inserted, perhaps working over part of the route only. The interval is chosen with regard to experience of passenger demand and to the planned round trip time of the route. In practice, the route structure itself may be adapted to obtain round trip times which suit the timetables.

Often, several routes form a forked structure sharing a common stretch, and it is desirable that they should together provide a regular service over that stretch. If all the prongs of the structure are the same length, this is logically equivalent to scheduling an individual route, but where lengths are different, the problem of integration on the common stretch can be considerable. Sometimes the planner has to choose between minimising vehicles requirements and providing a regular service.

The planner often has to experiment with many structures. Bus scheduling programs as described in the next section can be used to help determine the consequences of any particular stretegy, but it is also possible to have a computer automatically produce schedules covering several routes and providing regular intervals where possible on common stretches. Kwan (1982) has outlined such a system, and has since developed it further.

Bus scheduling

Proven programs exist for each of two rather different bus scheduling situations. We deal with these separately.

Route-based programs. When bus routes are scheduled individually, there is little room for optimisation. The main need is for a program which will speedily generate schedules, given some basic data, and will produce timetables and other working documentation.
The same is true to a certain extent for groups of routes sharing some common feature, for example operating along a single urban corridor.

Several programs have been written to deal with this problem. Some have already been mentioned, but of these, the only one now in use is the MINI-SCHEDULER, designed for the situation where timetables are developed according to demand. The system provides a set of commands whereby the user can specify patterns of work (these being descriptions of routes or variations of routes, with associated running times), and can build up the timetable by specifying interactively the times at which patterns are to operate. Once an acceptable timetable has been developed for each direction of travel, the user can ask the computer to allocate buses to trips, allowing deadheading if appropriate. The bus blocks generated are then each assigned to the optimal garage. The MINI-SCHEDULER is in use in a large number of North American companies.

TASC is a program designed for regular service scheduling. Originating in the University of Leeds, it is now part of the BUSMAN system, and is described by Hartley and Wren (1985). The user can define routes and their variants, and can specify regular or irregular services on these routes by giving the time of the first bus, a service interval to be operated until the next time given, a further service interval, and so on. TASC forms a schedule automatically, inserting deadheading trips (or additional service trips if appropriate) to balance arrivals and departures at individual points. It can deal with many garages, and arranges the schedule so that wherever possible buses start and finish their work near the same garage, before allocating garages optimally.

Once a schedule has been formed, an interactive amendment facility enables the user to make a wide range of adjustments. TASC maintains a database of the entire bus schedule, and the amendment facility can be used either to incorporate schedulers' special requriements before finalisation, or for later day to day adjustments and minor permanent alterations which do not warrant complete rescheduling. TASC is in regular use in the UK, Asia and Australia.

Both the MINI-SCHEDULER and TASC produce a wide range of documents for operational purposes which can be tailored to the needs of individual clients. In the case of TASC these include: public timetables (which can be printed automatically); working timetables showing each journey together with the notional vehicle number; running boards (detailed descriptions of the work of each bus); details of the scheduled mileage for each route, broken down if required by county or other administrative districts; working documents for recording the actual service operated.

Network-based programs. Often it is best to schedule a complete network, or a large part of a network, as a single job. This applies if some routes have irregular workings, if special journeys are super-imposed on regular routes, if the peak flow is directional and it may be better to run a vehicle dead to another route rather than back to its starting point, or if the time of peak demand varies according to route.

An American program widely used for both route-based and network-based bus scheduling problems, and for driver scheduling was RUCUS (see Nussbaum et al., 1975). RUCUS treated bus scheduling as a problem of network flow. The nodes of the network are departures and arrivals of trips. The arcs are scheduled journeys, together with selections of possible links between journeys at terminals and of some potential deadheading trips. Unfortunately, while algorithms are available which find the optimal solution to a network, it is necessary when forming the network to restrict severely the number of deadheading trips considered. While this may not be critical in scheduling individual routes, useful possibilities may be ignored when the method is applied to a large network. Schmidt and Knight (1981) have stated that the original vehicle scheduling component of RUCUS proved unsatisfactory in quality of results.

Although a new RUCUS II has been presented by Luedtke (1985), the bus scheduling algorithm would appear to be unaltered. Keaveny and Burbeck (1981) indicate that a similar method is used in the MINI-SCHEDULER, and neither of these programs is therefore suitable for network-based problems. TASC also does not seek the best solution when a large number of deadheading possibilities exist.

VAMPIRES (outlined by Hartley and Wren, 1985) is a network-based program adapted from an earlier railway locomotive scheduling program. It is based on a powerful heuristic, and although optimality cannot be guaranteed, the basic algorithm has never been bettered in extensive tests, and has often found solutions considerably better than those found by other methods.

A unique feature of VAMPIRES is that one can set a target number of buses. The program will use a lower number if possible, but where the target cannot be achieved directly, it will highlight those journeys whose alteration will make the target realisable. Often adjustment of a time by a minute or two is sufficient, but sometimes critical journeys may be curtailed or cancelled.

The first application of VAMPIRES to a real problem, in 1971, led to a schedule using 41 buses beign recast with 39, without any adjustment of journeys. Progressive reduction of the target from 38 to 31 resulted in management being shown large numbers of critical journeys whose revision they were able to accomplish, thus yielding a saving of ten vehicles. In many cases these revisions were simple retimings by a few minutes.

VAMPIRES is used regularly, and is integrated in the Wootton Jeffreys BUSMAN system in conjunction with TASC. While VAMPIRES can still be run on its own to solve specific problems, integration with TASC in cases where many journeys are of a regular nature has proved very successful.

In the integrated mode, TASC is used first on a number of routes or route groups separately; the scheduler adjusts each TASC solution interactively if he wishes, and the whole set of solutions is passed to VAMPIRES together with details of any additional journeys on non-standard routes. VAMPIRES will maintain the TASC schedule unless positive improvements can be made, thus ensuring tht the individual solutions already inspected and possibly adjusted by the scheduler are retained where sensible. Finally, the VAMPIRES schedule is returned to TASC for further interactive amendment and for document production.

More recently, a MICROBUSMAN system has been developed on IBM PC and compatible micro-computers (see Wren et al. 1985a). This combines an exceptionally user-friendly interactive database input with the VAMPIRES scheduling system, a new interactive amendment process, and the TASC output facilities. It is now used by several British companies.

Crew scheduling

Known as run-cutting in North America, crew scheduling is the allocation of drivers or of two-person crews to the bus schedule. On any bus route there is normally an agreed point for changing crews; sometimes there may be more than one. The work of any bus throughout the day may be split into segments bounded by relief opportunities, the times when the bus passes the change-over points. Depending on the nature of the bus route, the segments may vary from a few minutes to a few hours.

Scheduling bus crews consists of allocating a notional crew to each segment in such a way that all the segments allocated to one crew form a duty which satisfies a set of local rules. In practice, these rules vary widely from one organisation to another, even within a single country.

A duty working on only two buses is generally the most efficient where a meal break is compulsory, since no time is wasted changing vehicles, but it may not easily be achieved. A major difference between scheduling bus crews and scheduling personnel working at fixed locations arises because the bus may not be near a change-over point at the ideal time. As a consequence, the scheduler generally must examine many possible solutions before determining the best, and sometimes has to accept a solution without knowing whether it can be bettered. Often he must transfer a driver between vehicles several times, simply to make up a duty with sufficient work.

If meal breaks are not compulsory, scheduling is eased, since one or two single duties of perhaps seven to nine hours can be cut from many of the bus workings. This leaves scraps which must be fitted together efficiently, and many computer programs have been written to deal with this situation. However, in most developed societies, with notable exceptions (for example in some North American situations), meal breaks are required with consequent difficulties in both manual and computer scheduling. In fact, scheduling becomes steadily more complex as the maximum time worked without a break is reduced. While in some countries, six hours or more is acceptable, European Economic Community rules normally limit this time to five and a half hours, and local agreements in the United Kingdom often reduce it to four and a half.

It has long been known that bus crew scheduling was in theory an integer linear programming (ILP) problem. The number of variables and constraints is, however, such that practical solutions were originally possible only in simple situations. Most early attempts at crew scheduling therefore used heuristics, while more recently interactive facilities have been combined with these. However, heuristics have largely been abandonned because of difficulties in adjusting them to changed conditions. In parallel with the addition of interactive facilities, powerful ILP methods have now been developed and have proved highly successful. (2)

Heuristic approaches. Early heuristics for crew scheduling included those of Elias (1966), which by the early 1970s had been extended to solve satisfactorily a range of trial problems. Elias's method bore no resemblance to that of human schedulers. Wren's (1969) method, by contrast, was developed over a period in such a way that it grew to resemble a formalisation of processes followed by human schedulers, followed by a series of refining routines. This was reported by Parker and Smith (1981).

The program, known latterly as TRACS, was applied to many practical problems, producing satisfactory results in most cases, some of the schedules being put into operation and yielding savings.

Although many British bus companies sponsored these successful demonstrations of TRACS, none commissioned its full installation, partly because of inability of data processing management to come to terms with a system whose concepts they did not understand, and partly because of a fear that if the scheduling rules were to change, significant work might be needed to adjust the heuristics.

The RUCUS run-cutting system (see Nussbaum et al., 1975) was similar to TRACS in many ways, although its scope appears to have been narrower. For example, TRACS could create duties working on three or more buses, while RUCUS duties were restricted to two. RUCUS also appears to have suffered from difficulties in adapting to new situations. Schmidt and Knight (1981) reported that it was not adequately user-oriented and that the results must often be modified extensively by hand. Despite this, RUCUS was widely implemented in the United States, for example, as described by Landis (1981).

Reasons for the relative success of RUCUS as opposed to TRACS are varied. There was a greater willingness among American bus companies than among British to introduce new computer methodology, and implementations in the United States were assisted by substantial federal funds. American crew scheduling conditions are in general slacker than those in the UK, and it is easier for heuristic systems to adapt to a range of slack conditions than to a range of tight ones; the amount of adjustment needed for new users may therefore have been less in America. Further, duties working on more than two buses are relatively rare in America, but common in the United Kingdom, and it was often in the heuristics dealing with these that TRACS had to be adjusted from one organisation to another.

Interactive approaches. Because of the limitations of heuristic approaches, several bodies have developed interactive programs for crew scheduling, often containing elements of the heuristic tools.

RUCUS II, as described by Luedtke (1985), incorporates heuristics similar to those of RUCUS, but selected interactively. Luedtke does not state whether duties may also be created interactively.

Hildyard and Wallis (1981) outline an interactive runcutting system now marketed as the SAGE RUNCUTTER and widely used in North America. This has RUCUS-like heuristics driven by the user. Duties formed by these are displayed on the terminal, and the user may accept them or command the computer to try again with changed controlling parameters. The user may also specify duties completely for incorporation within the schedule.

The COMPACS program outlined by Wren et al. (1985b) is part of BUSMAN and incorporates many TRACS heuristics. The user may call particular heuristics and accept or reject duties proposed by the computer. He can form duties interactively and seek the aid of the program, for example by specifying part of a duty had having possible further parts displayed for his choice. The computer tells the user if he tries to specify a duty containing work already allocated. It will also tell him if he attempts to break any rules, but the user may then over-rule the computer.

Both COMPACS and RUNCUTTER accept input from their associated bus scheduling systems and display a wide range of information to help the user build up a duty schedule. (3)

Integer linear programming (ILP) systems. While interactive systems are attractive to the scheduler, being under his complete control, they do require a significant time for thought and interaction. This may be unacceptable in a busy scheduling office, or where the ability of the scheduler is in doubt. There is therefore a need for good fully automatic systems to compile large schedules speedily, either for implementation or so that the consequence of variations in operating patterns or working conditions may be assessed.

ILP provides such systems, of which the most widely used is probably HASTUS (see Rousseau et al. 1985). HASTUS, which was developed in the University of Montreal and is marketed by GIRO, comprises four stages.

HASTUS-MACRO, the initial stage, forms very many idealised duties. (An idealised duty is one whose starting, finishing, and meal break times are known, and occur at multiples of perhaps fifteen or thirty minutes; no actual vehicles are allocated to the duty). HASTUS-MACRO uses ILP to select a sub-set of these duties which together provide at least as many drivers as there are buses at any time of day, at minimum cost. These idealised duties are assumed to be approximations to the actual duties which will be created.

The second stage partitions the actual bus work into stretches which match as closely as possible the idealised duty halves proposed by HASTUS-MACRO. This partitioning is done by heuristics.

Next, pairs of actual stretches are matched optimally to form complete duties. Rousseau et al. state that when duties with more than two stretches of work have to be considered, combinations of smaller stretches are created before using the matching algorithm. The present author, having worked on computer scheduling of drivers for nearly twenty years, believes that the decision as to how to combine three or more stretches into a duty depends on the duty as a whole, and not on pairs of stretches. Further, the best stretches for combination are unlikely to emerge from the second stage process which considers only idealised two-stretch duties. It may be that in North America the importance of forming good duties with more than two stretches is less than in some countries. Certainly, the author has met several situations in the United Kingdom where finding a unique duty with three stretches has been essential in obtaining a schedule with the minimum number of drivers.

The final stage re-examines the partitioning of each bus in turn; stretches are re-matched if this gives an improvement.

Significant savings using HASTUS in Montreal have been claimed by Rousseau and Blais (1985). HASTUS is in use or currently being installed in several countries of the world.

An entirely different approach is adopted by IMPACS*, developed in the University of Leeds (see Wren et al. 1985b). In theory, it is possible to generate every conceivable crew duty and to formulate an ILP model which would ensure that a selection of duties were obtained covering every segment of bus work at minimum cost. Additional constraints on numbers of particular types of duty can be added. The model used specifies that at least one driver is available for any bus at any time.

In practice, the numbers of possible duties and segments are generally too large to achieve a solution. IMPACS therefore uses processes which reduce these numbers without sacrificing the quality of the results.

---

* Not to be confused with an interactive system of the same name described by Howard and Moser (1985)

Following these, models are easily solved with about 4000 possible duties and over 300 segments. In most situations this allows problems of about eighty duties to be solved in a single pass, although larger problems have been solved on occasions.

Most bus crew scheduling problems in major cities involve many more than eighty duties. IMPACS therefore divides the bus schedule if necessary into a series of smaller units. Before forming these sub-problems, IMPACS analyses the bus work, identifying features likely to cause difficulties in forming an efficient duty schedule. It then finds particular pairs of relief opportunities on different buses whose use in an individual duty would overcome one of these difficulties. Each pair identifies two buses which ought to be treated together, and these are put in the same sub-problem. The sub-problems are solved sequentially, as outlined below, and a target amount of bus work comprising undesirable or inefficient duties from earlier solutions is carried forward and merged with the work to be solved in the next sub-problem.

When the final sub-problem has been solved, a complete schedule exists, consisting of all the duties in the final solution, plus those not carried forward from earlier ones. A refining proces is applied to this schedule, in which improvements are sought by exchanging stretches of work between duties or moving a particular change-over to the previous or next opportunity on the same bus.

Just as HASTUS cannot guarantee optimality after matching actual and ideal portions of bus work, so IMPACS cannot guarantee the optimality of the solution following the sub-division. However, the sub-problems have been carefully constructed, noting the possibility of certain duties being essential to a good schedule. In practice, no solution has ever been found by other means which used fewer crews than that found by IMPACS, and sometimes it has been possible to show by theoretical arguments that no such solution can exist.

The solution of a sub-problem, or complete smaller problem, consists of several stages. First, the bus schedule is examined, and certain unlikely relief opportunities are removed. This process effectively reduces the number of rows in the ILP.

In the second stage, a large number of duties (up to about 15,000) is generated, using the remaining relief opportunities. The algorithm ensures that all reasonable duties are considered, and is governed by parameters specifying the legality and quality of the duty. For example, one may stipulate that no early duty is to contain less than five hours driving time.

The third stage rejects any duty that is wholly included within another duty of the same cost. In the fourth, a duty is rejected if there are at least n more efficient duties covering each segment of bus work within it. A good value of n is often 50. Following these processes, the number of duties is reduced to around 4000.

Stage 5 is the application of a specialised ILP algorithm. This usually contains three phases, although they may be run automatically in sequence. First, the LP is solved without integer restrictions. Next, the number of duties is rounded up and imposed as a lower bound, while constraints governing the numbers of duties of particular types are added if required. Finally, penalties are added to the costs of undesirable duties, and a good solution is obtained by a branch and bound process.

Stage 6 is a small-scale application of the refining process used after the final sub-problem. This enables relief opportunities or duties which have previously been rejected to be reinstated.

Wren (1985) describes early applications of IMPACS. Regular users include London Buses Ltd. (the world's largest unitary bus undertaking), and it has produced savings elsewhere of up to ten percent in crew costs. It is now part of BUSMAN, where it can be used in conjunction with the interactive facilities of COMPACS.

There have been several comparative trials of IMPACS and HASTUS. In some cases, the results have not been made known, but in those known to the author, IMPACS has always produced the cheaper schedule. Greater Manchester Transport chose to implement IMPACS after it produced a solution with 193 duties (compared to 197 in the existing schedule) while HASTUS required over 200. It is believed that the better performance of IMPACS is because it has been designed for the British situation which is generally more complex than that in North America.

Several other ILP systems have been developed, including RAMCUTTER from SAGE (1985), of which no details have been made available. Suppliers of systems based on ILP sometimes claim that because of the processes used, the best possible solution is guaranteed. In fact, all ILP systems for solving large duty scheduling problems include simplifications which can detract from their optimality, and potential users should not be influenced by such claims.

Crew rostering

In some places, notably North America, drivers pick, in order of seniority, duties to be worked regularly. Micro-computer based software, for example by SAGE (1985), is available to assist in combining the duties and managing the picking process.

Elsewhere, drivers work rotating rosters so that ultimately every driver works every duty of the system or sub-system. The style of such rosters varies considerably, and it is therefore difficult to develop universal software. There are two distinct classes of rotating rosters. In each, duties are listed in rows, usually of a week's work, and drivers progress cyclically through the rows.

In the first class, a driver usually works the same duty every weekday, possibly combined with a Saturday or Sunday duty or both, with weekdays off in lieu, and the payment for a week is the sum of daily payments. The skill in creating such a roster lies in placing the days off, in determining the positions of different types of duty, in associating Saturday or Sunday duties with similar weekday duties, and in spreading special features evenly through the roster. A prototype automatic system for this class of problem, with interactive features, is described by Townsend (1985), and has been installed for London Buses Ltd.

The second class applies where payment for a week depends on the precise combination of duties. Here, in addition to the above, it is necessary to group five different duties to give a week's work so that the total cost of all weeks is minimised. A method for this, due to Bennett and Potts (1968), first assigns rest days, then finds a good allocation of duties to the remaining positions in the roster. The present author and colleagues extended this to a British problem in 1979, but management could not sufficiently specify the rostering rules. A rostering program has been used in Brussels for a very specialised problem since the late 1960s: an English language description was presented by Crouch and Morton (1970).

COST ESTIMATION

A good estimate of the cost of any bus service must be made during planning. Cost depends on individual duties, and critical features of a bus schedule leading to the need for extra drivers are often not discovered until these are formed. This makes a rough estimation subject to errors, although it may sometimes be possible to adjust the bus schedule in order to remove a critical feature if an actual crew schedule requires more duties than has been estimated.

Four programs already mentioned have been used to provide cost estimates of different degrees of accuracy, and a further program, CREWED, as yet unpublished, has recently been developed by the author in conjunction with the National Bus Company. These all start from a notion of the bus schedule and estimate the number of duties required. The estimation facilities of each are outlined here.

TASC incorporates a very fast process in which a few basic parameters of the labour agreement are used with the bus schedule it has already produced to assess the numbers of duties in each of five broad categories. This has obtained generally good results enabling the user to decide whether a trial schedule is worth developing.

HASTUS-MACRO is itself an estimating program with more comprehensive parameters than TASC, using an approximation to the bus schedule. Being based on ILP it probably uses much more computer time. It also gives an assessment of the numbers of duties of various types. Its use as an estimator has been described by Mitchell (1985) and Dupuis (1985).

CREWED uses parameters between those of TASC and HASTUS-MACRO in complexity, with an approximation to the bus schedule. It uses fast heuristics (generalisations of those in TRACS), and produces idealised duties with starting, finishing and meal break times. It shows when there are surplus crews available for extra work.

COMPACS incorporates a procedure based on the actual work rules and bus schedule. It produces total numbers of duties of each type. IMPACS produces an estimate of the number of duties required as the first phase in its fifth stage. This is very accurate in total numbers and cost, but uses significant computer time. One British company has, however, used this estimator to evaluate different work rules, proceeding to a full schedule in promising situations.

PERSPECTIVES ON THE USE OF COMPUTERS WITHIN THE BUS INDUSTRY

General management systems

Some suites, for example that from SAGE, combine scheduling with more general management functions such as timekeeping systems combining payroll with day to day driver assignments. There is also a full range of free-standing software to assist the management of the bus industry. Many systems have been developed in house, generally on micro-computers, while others are commercial products. UMTA (1985) contains a catalogue of the relevant micro-computer software developed in the US.

A survey of British bus companies carried out by Haywood and Trevette (1983) revealed that the most common systems then active were, in decreasing order: Payroll and superannuation; Budgetary control; Central stores control; Purchasing and supplies; Fuel and oil records; Ledgers. The most common projected systems were: Route revenue and costing; On-bus revenue control; Mileage recording; Maintenance job costing. It may be significant that at that time in the UK, few of the active systems were specific to the bus industry.

A study in Cardiff (UK) in 1979-80 identified five key areas where computer development could be of benefit: Reconciliation of revenue and analysis of waybills; Recording of bus operations and mileage analysis; Costing and analysis of engineering operations; Route costing and financial modelling; Stores management.

The first implementations after this study were in passenger and mileage statistics, with reports on performance of services. There followed a management information system for engineering operations, commercial fleet management packages being rejected as unsuitable for bus fleets. Recent information from Cardiff is that systems are working in all five key areas, aimed at all levels of management. This has been achieved without expensive hardware or outside help.

At the other end of the spectrum, the Southern California Rapid Transit District (Los Angeles) has commissioned a sophisticated mainframe-based integrated system, TRANSMIS, covering all aspects of operations and management. Work started in 1981, and is due for completion in 1986. The first phase, covering vehicle and material management, production control, financial management, and capital project control, was undertaken by Peat Marwick. The second was commissioned from Arthur Andersen in 1983, including planning, scheduling, transportation, payroll and personnel. It is estimated that over $50 million will be saved by TRANSMIS over a five-year period. TRANSMIS draws on the experience of several consultants and combines proprietory software with custom-written systems.

Micro-computer or mainframe?

The distinction between mainframe, mini-computer and micro-computer is increasingly blurred, but generally the first of these requires a controlled environment, the second can stand in an office, while the third can sit on a desk. Probably all the systems mentioned in this chapter can operate on a mini-computer, and many on a micro, although some users may prefer a larger machine able to handle more applications simultaneously. Such larger machine may be in a systems department, whose staff, if of appropriate quality, can assist in installation and operation of systems.

The advent of the micro-computer can bring IT nearer the individual manager at any level in the bus industry. Although desk-top computer terminals have been available for many years, and can appear to the user similar to micro-computers, practical (often political) factors have prevented their full exploitation.

The traditional mainframe in the bus industry has been managed by a DP group reared on routine commercial systems. The DP manager has distrusted specialist transport software whose use he does not understand and whose algorithms are beyond his experience. He has therefore sought to exclude systems written by outside consultants, using delaying tactics, or persuading management that it would be better that his staff should develop equivalent systems of their own. These equivalent systems have then failed to materialise.

Often the mainframe has not been exclusive to the bus management, being under the control of a city treasurer. In such circumstances, bus management has generally felt that its urgent problems were being made subservient to the routine work of a finance department.

However, a mainframe can work for the bus executive, provided that the priorities of access are determined according to the needs of the bus business, and that systems staff is willing to explore transport applications. Indeed, in a large organisation, a mainframe will probably be essential for several years to come.

Smaller companies have already properly turned to micro-computers, as instanced by Cardiff, where it is felt that there is no ideal central point for information, but rather an information ring. Individual managers will have their own micro, or access to a micro, depending on their functional responsibilities. The micros will be linked, either physically, or by transfer of floppy discs. In this way, the individual machine is under the control of a single section of the organisation, and is used for that section's functions, while relevant information is passed to other sections. (4)

Summary

Many systems have been developed for the bus industry. While the exciting problems and easily quantifiable savings have been found in scheduling and network planning, many organisations are reaping the benefit of the automation of quite simple processes. Much competing software is available in some fields, and the potential user is advised to examine carefully the relative quality of results.

It is not advisable for a bus company to write its own scheduling or network planning systems, as much effort has been invested in those commercially available. However, companies have successfully written softare for various aspects of management information, and this option can provide systems tailored to their own needs.

REFERENCES

Belletti, R, Davini, A, Carraresi, P and Gallo, G. (1985). BDROP: a package for the bus drivers' rostering problem. In: Rousseau (1985), pp.319-324.
Bennett, B.T. and Potts, R.B. (1968). Rotating roster for a transit system, Transportation Science 2, pp.14-
Bolland, J.D. (1978). Simulation of a bus network based on a multipath passenger assignment algorithm. Leeds University Ph.D.
Brandani, V., Cataoli, G. Orsi, G. and Toni, P. (1981). Bus scheduling program development for A.T.A.F. Florence. In: Wren (1981), pp.71-83.
Ceder, A. and Stern, H.I. (1985). The vehicle trip procedure used in the Autobus vehicle scheduler. In: Rousseau (1985), pp.371-390.
Crouch, F.O. and Morton, N. (1970). An example of computeriosed duty rosters in operation in a large mixed fleet. In: Proc. Symposium Tomorrow's Bus World, University of Newcastle upon Tyne, pp.15-25.
Dupuis, D. (1985). Automatic crew scheduling: new operating management and service opportunities. In: Rousseau (1985), pp.145-148.
Elias, S.E.G. (1966). A mathematical model for optimising the assignment of man and machine in public transit run cutting. West Virginia University Engineering Experiment Station Research Bulletin 81, 62 pages
Elias, S.E.G. and Smith, N.S. (1970). The development and demonstration of an automatic passenger counter. West Virginia University Engineering Experiment Station Bulletin 94 (2 volumes).
Hartley, T. and Wren, A. (1985). Two complementary bus scheduling programs. In: Rousseau (1985), pp.345-369.
Haywood, P.J. and Trevette, I. (1983). Progress in bus operators' systems development. In* Leeds (1983).
Hildyard, P.M. and Wallis, H.V. (1981). Advances in computer assisted runcutting in North America. In: Wren (1981), pp.183-192
Hoffstadt, J. (1981). Computerized vehicle and driver scheduling for the Hamburger Hochbahn AG. In: Wren (1981), pp.35-52.
Howard, S.M. and Moser, P.I. (1985). IMPACS: a hybrid interactive approach to computerised crew scheduling. In: Rousseau (1985), pp.211-221.
Keaveny, I.T. and Burbeck, S. (1981). Automating trip schedulig and optimal vehicle assignments. In: Wren (1981). pp.125-145.
Kwan, R.S.K. (1982). A simple aid to headway integration. In: Leeds (1982).
Lampkin, W. and Saalmans, P.D. (1967). The design of routes, service frequencies and schedules for a municipal bus undertaking. Opl. Res. Q. 18, 375-397.
Landis, M.G. (1981). A perspective on automated bus operator scheduling: five years' experience in Portland, Oregon. In: Wren (1981), pp.61-67.
Leeds (1969-1985). Proc. Annual Seminars on Public Transport Operations Research. Dept. of Adult and Continuing Education, University of Leeds.
Leprince, M. and Mertens, W. (1985). vehicle and crew scheduling at the STIB. In: Rosseau (1985), pp.149-178.
Li Yihua (1985). The application of the microcomputer in bus and crew scheduling in Shanghai. In: Rousseau (1985), pp.179-198.
Luedtke, L. (1985). RUCUS II: a review of system capabilities. In: Rosseau (1985), pp.61-116.
Mitchell, R. (1985). Results and experiences of calibrating HASTUS-MACRO for work rule cost at the Southern California Rapid Transit District, Los Angeles. In: Rousseau (1985), pp.119-136.

Mitra, G. and Darby-Dowman, K. (1985). CRU-SCHED; a computer based bus crew scheduling system using integer programming. In: Rousseau (1985), pp.223-232.

MVA (1985). VIPS interactive planning system for public transport. The MVA Consultancy, Woking.

Orford, K.J. (1974). The bus routing project in Coventry. In: Leeds (1974).

Parker, M.E. and Smith, B.M. (1981). Two approaches to computer crew scheduling. In: Wren (1981), pp.193-221.

Piccione, C., Cherici, A., Bielli, M. and La Bella, A. (1981). Practical aspects in automatic crew scheduling. In: Wren (1981), pp.223-235.

Rousseau, J-M. (Ed.), (1985). Computer scheduling of public transport 2. North-Holland, Amsterdam.

Rousseau, J-M. and Blais, J-Y. (1985). HASTUS: an interactive system for buses and crew scheduling. In: Rousseau (1985), pp.45-60.

Rousseau, J-M, Lessard, R. and Blais, J-Y. (1985). Enhancements to the HASTUS crew scheduling algorithm. In: Rousseau (1985), pp.295-310.

SAGE (1985). SAGEPAC: the microcomputer based product line from SAGE. SAGE Management Systems, Toronto.

Schmidt, J.W. and Knight, R. (1981). The status of computer-aided scheduling in North America. In: Wren (1981), pp.17-22.

Townsend, W. (1985). Bus crew rostering by computer. Leeds University M.Sc.

UMTA (1985). Microcomputers in transportation: software and source book. National Technical Information Service, Springfield, Va.

Williamson, R.H. (1985). BUSMAN: the United Kingdom's integrated approach to transit scheduling. In: Rousseau (1985), pp.19-43.

Wren, A. (1969). Applications of computers to transport scheduling in the United Kingdom. West Virginia University Engineering Experiment Station Bulletin 91, 77 pages.

Wren, A. (1971). Computers in transport planning and operation. Ian Allan, London, 152 pages.

Wren, A. (Ed.), (1981). Computer scheduling of public transport. Papers based on presentations at the International Workshop held at the University of Leeds, 16-18 July, 1980. North-Holland, Amsterdam.

Wren, A. (1985). Recent developments and applications of computer scheduling. In: Leeds (1985).

Wren A., Foster, E. and Hartley, T. (1985a). Scheduling using microcomputers. In: Leeds (1985).

Wren, A., Smith, B.M. and Miller, A.J. (1985b). Complementary approaches to crew scheduling. In* Rousseau (1985), pp.263-278.

EDITORS' NOTES

(1) It is interesting to compare the approach to the scheduling problem adopted in the rail industry with that described here for the bus industry. For example, Chapter 7 suggests that, within British Rail, the timetabling of services traditionally precedes the detailed scheduling of rolling stock, whereas the systems described in the current chapter treat timetabling as an integral part of the scheduling problem. There are other differences between the approaches adopted reflecting subtle differences between the problems faced.

(2) It is interesting to contrast this conclusion with the view put forward in Chapter 9, where it is argued that exact optimisation methods are impractical for the road freight routeing problem.

(3) For a discussion of the arguments for human intervention in the scheduling process and particularly the role of computer graphics in this, see Chapter 11.

(4) The advantage of locally networked microcomputers are mentioned in several chapters of this book (see, for example, Chapter 7 on their role in rail operations and Chapter 12 on their role in the travel industry).

Information Technology Applications in Transport, pp. 235–255
P. Bonsall, M. Bell (Editors)

Chapter 11

INTERACTIVE GRAPHICS FOR ROUTEING AND SCHEDULING

Guy Lapalme and Michael Cormier
University of Montréal, Canada

VARIATIONS IN THE PROBLEMS OF ROUTE PLANNING

The problems of route planning can be divided into three groups routing, scheduling, routing and scheduling. Bodin et al. (1983) (1) provide this classification of route planning problems, give a coherent formulation for each of them, describe solution techniques both exact and heuristic but only give hints about the use of interactive graphic approaches. This paper shows how the use of sophisticated color graphic computer systems can help a route designer or dispatcher to improve on solutions obtained by a system implementing operations research routing techniques.

These solutions are almost always heuristic ones because no efficient algorithms have yet been developed for the sizes required by practical problems. Solving a routing and scheduling problem involves many compromises within a set of conflicting goals and constraints, e.g.: shortest route vs the use of main roads, better average service rather than a better overall service.

Consequently, the results must be evaluated by hand. This process usually prompts some more modifications who then in turn must be evaluated.

A routing and scheduling system deals with a collection of entities requiring service. Many attributes can be associated to an entity: its load, its requirements, times for service either definitive or within a time window giving an interval of time during which service has to be given. Relations of precedence can also exist between entities: for example, a pickup activity of an entity must occur before its delivery.

If the entities to be serviced have no temporal restrictions and no precedence relations between them, then we have a routing problem. If each entity has a definitive service time, then we have a scheduling problem. But "pure" problems are rare and most often we have a combination of routing and scheduling. This does not mean that the techniques for solving pure routing or scheduling are not useful, in fact, in many problems these aspects can be dealt with separately or given more or less emphasis according to the tightness of constraints or to the relative costs they imply.

STEPS IN ROUTE PLANNING

The sensitive spot in route planning is the design of algorithms to solve and optimize specific problems. However, many other activities requiring computer treatment are also implied in the whole process of route planning. First of all, the "universe" defining the problem must be described in a suitable form for the algorithms, in particular the location of the clients, the distances between each of them, the fleet of vehicles, etc. When all the data is collected and when the objectives are well defined, then an appropriate algorithm is run. Once the routes are produced, they are evaluated to check if the algorithm is working well on that set of data and if the routes meet the basic objectives. Sometimes a few modifications are required to adapt a computerized solution to the constraints of real life specially those that cannot be modelled. Finally, when the routes are judged satisfactory, they can be implemented.

Defining the Problem

The first steps in a route planning problem are to locate the entities (points of pickup and delivery) and to determine the time of service or their time-windows. In a routing problem, the geographic situation of the entities is important; also useful are their relative positions (i.e.: the distance between them, the available paths, natural barriers blocking direct access from one point to the next, etc.). One intuitive and very effective way to visualize a problem is to plot those entities on a road map. The coordinates of the entities can be used inside the computer to evaluate the relative distances between them by finding the euclidian distances (as the crow flies). However, a real road network does not allow going straight from one point to all the other points: usually one has to go past intermediate nodes before reaching a destination.

A better evaluation of the distances can be computed given that the road network is in computer readable form. This implies that within that framework, we can find the coordinates of each street corner and that we know which corners are connected to each other. So we have a network whose nodes are street corners and whose arcs are the street blocks.

To find the shortest distance between two points in this network, we can use the classical algorithms for finding shortest paths, (see Denardo and Fox (1979) and Dial et al. (1979)). The distance itself is seldom needed, but what is sought is the duration of time to go from one point to another: in that case, the distance is divided by a mean speed to find the time duration. Of course, many variations are possible for mapping distance to time: e.g. having different functions depending upon the distance, taking into account when the path goes in a densely populated area or on a highway.

Having a computer managed network gives access to a whole array of automatic techniques for computing other important values needed by the algorithms. Maps of the networks with routes can also be generated.

There is one drawback, the clients have to be located on this same network, i.e. pinning down the node in the network which is closest to the client. As Table 1 shows, a real network comprises thousands of nodes, so finding the right one is not always easy. A program has to be developed to locate the clients given their address, postal code or any other available information. We may also imagine that in an interactive session, a map would be drawn on the screen and that the operator would locate the client with a pointing device (a mouse, a light pen or a joystick); the computer could then find the appropriate node to locate the client.

Once a client is located, other information can also be tagged such as the load, the time slot for pickup or delivery, the name of the client, etc.

Other important informations are the costs. The operating costs associated with vehicles can be difficult to estimate exactly. So these are replaced by other objectives like distance travelled, number of vehicles. Crew costs are usually determined by union agreements but they often imply a very complicated cost structure which has to be simplified in the planning of routes.

Finding an Initial Solution

Once all the information is available, one can ask for a solution by routing/scheduling algorithms (see Bodin et al. (1983) for their description). They are usually heuristics that give good solutions but not "optimal" ones. As with many other O.R. problems, the routing/scheduling scenario has to be dramatically simplified before going for solution in the machine. Many constraints, whose mathematical formulation would be difficult, have to be discarded (for example: quality of service, no left turns, etc.). Some constraints can be violated but not too often as long as big savings are achieved (e.g. we might accept to be a little late at a client because doing so enables the pickup of other clients with the same vehicle; starting a new one usually implies a very large fixed cost).

Table 1 - Sizes of typical road networks

Name	Population	Area in km^2	# arcs	# nodes	# streets
Toronto (Metropolitan Toronto)	2 500 000	663	31 650	21 570	8 709
Montreal (Island of Montreal)	2 000 000	488	29 508	19 258	6 063
Sherbrooke (City and suburbs	100 000	224	3 247	2 380	1 297

Evaluating the Solution

Given a set of constraints, computers can generate many possible route schemes. As we have pointed out before, those algorithms are heuristic and cannot take into account all the hidden constraints or habits that the dispatcher has to take care of. Thus, before implementing a set of routes, it is mandatory to scrutinize them to detect any problems; usually, drivers or dispatchers are too happy to point out a route that does strange turns or forgets to turn at the right place. Those errors are usually the result of input or network errors (i.e.: two streets that meet in reality do not touch each other in the network or vice versa).

The best way to appraise such routes is to draw them on a road map; usually many of them are on a single territory and we are interested in having a more global improvement than only one route at a time; to differentiate them, we give a distinct color to each route. Handdrawing those results is very slow and tedious, so the generated routes are accepted as they are or worse they are rejected as a whole. What is needed is an automatic draughting tool to display both the road network and the routes in distinct colors. Preferably, this should be a high resolution system for a drawing to appear as it would on a real map. For that purpose, two major kinds of hardware are available: color plotters and color graphic screens. The firsts give a very good resolution and beautiful drawings but do not allow easy interactions to achieve the fast modifications that we want to implement. Interactive graphic screen systems are another possibility: networks and routes can be drawn quickly and the drawings can be selectively erased and redrawn to reflect any updates that the user might feel necessary. Of course, at the end of this evaluation process it will be useful to have the possibility of keeping a hard copy of the final routes either through a printer, a plotter or photographs. (2)

Modifying the Solution

As we said in the previous section, routes produced by standard algorithms are "almost" right but small modifications have to be made to the routes.

This can be done by editing the list of nodes describing a route but this process is very tiresome and error prone. Just keeping the consistency of the data base is much work, even if we are not taking into account any optimization aspect. We recommend the use of specialized route editors to display and update the routes safely, the effects of the changes being displayed immediately. We have built such tools, the next section describes them in more detail.

Implementing the Solution

Once the routes are judged satisfactory, perhaps after many iterations between the evaluation and modification steps, they can be implemented. The computer can help in preparing detailed worksheets giving the exact path to follow between each entity by listing all streets occurring in the shortest paths going from one entity to another.

The computer can also be useful to keep track of the current status of routes as they are being done, noting any discrepancies between the planned and the actual routes. This implies that we have a real time computer system and a mean of inputting regularly the actual positions of the vehicles. One can imagine an automatic location tool using a transmitter on the vehicle picking up a signal from a tower at a known position in the city when it is in its vicinity and sending this position to the central computer. By having many of those towers, it would be possible to get a very good approximation of the location of each vehicle at any time. But this infrastructure is quite costly and implies a whole gear of electronic equipment which could not probably be justified for one application only. This system should be shared by many organizations. Right now, a system of this sort is known to be in operation in open pit mines where distances to look over are very great and a control is necessary for achieving long range goals of productivity, more details can be found in the Canadian Mining Journal (1975).

We have seen that interactive graphic computer systems can help in all the steps involved in the planning and implementation of routes. The clients and the resources can be more easily entered, checked immediately and modified if we have an interactive computer system. As the information always have a geographic component, it is very natural to use graphic tools to represent the location of those entities. The evaluation of alternative solutions can also be eased by using an interactive graphic computer system because the human eye is very quick at detecting irregularities and discrepancies on a route. Using a graphic input module (lightpen, mouse or cursor) to indicate the modifications results in a very fast and easy to use system.

CASE STUDIES

From the different activities involved in route planning, two tasks are directly related to the use of interactive graphic techniques: the evaluation of the solution has to be done using routes drawn on a map and the solution should be easily modified. This leads to an automatic tool for displaying the routes on a map of the surrounding area in order to tell whether the routes are appropriate or not. With such a tool the whole operation of evaluation is speeded up because graphics are using the ability of the human eye to interpret spatial information. Even if we are using quite sophisticated algorithms, it is almost certain that the routes generated will not be 100% satisfactory and that a few modifications will be required. Some constraints of the problem had to be ignored or neglected due to time and memory limitations on computers; special cases often appear in real problems. However, an experienced person (a manager or dispatcher) can easily identify many of the weaknesses of the routes and tell how they should be improved. Not only more sophisticated algorithms or powerful machines are needed, but also an interactive tool to modify the computerized solution and make it more appropriate in a given context. Moreover using such a tool has a psychological impact; in fact, people who are reluctant to accept routes established by a black box algorithm will more likely adopt them if they have been involved in their elaboration.

The use of interactive graphics in the development of route planning systems is something recent. Nevertheless some work has been accomplished in that field. Cullen et al. (1982) and Fisher et al. (1983) use color graphic microcomputers mainly for the display, the main computations being made on a remote mainframe linked by a telephone line. Babin et al. (1982) features a transit planning system running on a stand alone powerful microcomputer. It aims more at displaying analytical results in form of tables and graphs than at showing routes over a network; this system also includes an interactive network editing module but not a route editing one. Tuan (1984) developed a system for the assignment of buses in a dial-a-ride context. This system allows dispatchers to enter the position of the clients, visualize routes, add new clients to routes and obtain information about the routes; however the dispatcher still has to decide where to add the new clients, the computer is not used to analyse the situations and give him suggestions.

For the last four years, our team at the Centre de recherche sur les transports of the Université de Montréal has invested a lot of effort in the development of interactive graphic tools. To help us in the task of locating clients on the networks we developed an interactive tool that can do searches efficiently and solves conflicts by asking the user for more precision. Another experience from which we learned a lot was the writing of an editor for routes for the transportation of students to school: the routes are first generated by a set of programs based on heuristics, then they are visualized and modified by the editor and finally they are given as input to a program that tries to minimize the number of vehicles required: the editor is only a part of a production system. Then we began the development of an editor for routes for the transportation of handicapped persons. Graphically it is a more demanding task since time information has to be displayed on top of the spatial information. Finally our most recent project is the design and development of a complete tool for the dispatching of vehicles in the context of pickup and delivery of goods. By complete tool, we mean software that will help the dispatcher with all his tasks (location of clients, suggestions for inserting a new request in a route, edition of routes, ...).

Right now those programs are at different stages of development and working on different computer systems but we are in the process of designing a whole new system starting from our many experiences and, we must admit, the many false tracks we ran into in this completely new ground of interactive computer graphic systems. Those systems can be seen as steps toward an integrated computer graphic system which could address all aspects of route planning and implementation.

Before describing these systems, we give the basic data sets which are common to all of them.

The main one is the geographic data base of the road network: it gives the coordinates of each street corner, the range of addresses on each block, mean speed on that section of road, any turn or direction limitations (e.g.: no left turn or one-way street), etc.

For certain applications, for example long distance transportation, one does not need, or want, so much detail but one should settle instead for an aggregate network giving only the main roads or routes and being satisfied with approximate distances within cities. The level of detail to be kept should be in accordance with the costs or the time for each section of road; for example, if a truck spends 95% of its time on the main roads, we could easily accept the exclusion of small streets in the city, likewise if we are dealing with delivery in the center of the city then we should have a much more detailed network.

One should not strive for precision for its own sake because this implies a bigger network in terms of more nodes and arcs and this alone brings new costs: it will need more disk and memory space to keep the network, it will take more time to process and also it will mean much more work to maintain it since there is more information to start with.

One other data file common to all systems is the entities file giving the list of clients and/or resources with the coordinates or some other means of locating them on the network. Instead of coordinates, node identification numbers are used; they can usually be found more rapidly than the coordinates themselves, and are easier to handle by algorithms.

Location

This process is an essential and an often neglected part of route planning. This step can also be a very time consuming one if the right tools are not given.

All algorithms and programs usually assume that the position of each entity on the network is known. This position has to be known in terms of the network as it is stored in the computer. This is usually in the form of universal coordinates or arbitrary node numbers. It would be inconceivable to ask a client to give us his "universal coordinates" before inputting his demand in our system, so we have to find a fast way of going from easily accessible information, e.g. addresses or postal codes, to computer oriented information.

This process is conceptually simple because in our network we are given the range of addresses for a street block. It is therefore only a matter of finding the right street block. Unfortunately, it is not always simple because many complications can arise: the first one is that too often the address sought is not a street corner so we need to extrapolate to locate the entity; another problem comes from the fact that network data files are often incomplete and addresses are not necessarily given at each corner; just finding the right street name can sometimes be difficult because of street name changes, badly written names (Young instead of Yonge), ambiguous names (there might be more than one "Church Street" in an urban region). It might happen that one name is used for many kinds of road (for example, St. Peter Street, St. Peter Road, St. Peter Crescent, St. Peter Place, etc.).

Of course, we could insist on having all this information right in the first place but this implies a very long and tiresome process.

We have developed an interactive location program on an IBM PC/AT with a 20 megabyte hard disk: the program displays a form with fields to be filled according to the information available.

Values in the fields are validated to eliminate erroneous characters (for example letters instead of numbers, non alphanumeric characters, etc.). Once the fields are filled, the computer searches the database to find the location of a given address. If the address can be located immediately, then the information is displayed and recorded in a file; if the information given does not determine an exact location then various heuristic rules have been developed. We first try to locate the right street, either by giving the names of streets that appear before and after the name sought; the exact name can usually be found by looking at the surrounding names. Once the right street is found, we give the addresses occurring at each street corner and the name of the crossing street. So if the address cannot be found exactly but the name of a crossing street is known then the entity can be located on a nearby node.

Even if this system has been implemented on a microcomputer, it is able to handle quite large databases. For example, response time is almost instantaneous even for a partially specified address within a city like Montreal or Toronto (more than 6 000 streets and 19 000 nodes for each network). This tool is aimed at persons receiving calls in real time and interacting directly with the client if he cannot be located immediately.

Another feature which has proved to be very useful is the ability to give names to addresses; once an address has been located, we can give it a name (it could be the name of the building or of the client) and use this name the next time to refer to it. This naming scheme is very useful and is the fastest way to locate clients: the entities are often located at the same points, the warehouses are always at the same places, and regular clients do not move often.

This interactive system is a great time saver, much better than our previous batch system where a set of addresses was run against the database and a list of rejects was given at the end of the run. Those rejects had to be corrected or specified more precisely. This iterative process can be quickly improved by having an interactive dialog with the user at the moment of input when the information is still fresh and available.

Interactive Route Editor

We now describe a tool to evaluate and modify routes. The overall scenario of an editing session is the following: display commands are used to show the routes; modification commands allow changes in the routes which are then stored.

Display commands. With these commands we can display the whole network or just enough to show the chosen routes, the stops can be linked either with the shortest path on the network or as the crow flies. Figure 1 shows five routes over the network of Sherbrooke, stops are indicated by dots linked by the shortest path and each route is in a distinct color.

Once the routes are displayed, we may need to clear up the situation for a better evaluation: for example, by changing colors of the routes, by removing a few from the display or even zooming in on a particular area of interest. Once we have a satisfactory display, we can start the modifications themselves.

Modification commands. We have three levels of modification:

- route level: we can create or delete new routes.
- stop level: we can add and remove stops from a route; we can also reorder stops within a route and move stops from one route to another. When the modifications are made the result is immediately reflected on the screen.
- within a stop: a number of attributes characterize a stop; for example, the quantity of goods to pick up or deliver, the time window for servicing, the names of the people to pick up, etc. Updating that information does not involve 2-D spatial information and can be done with the usual text editing functions. But care must be taken to update the global attributes of the route accordingly: for example, total quantity, total time length, total number of people, etc.

It is very important that the modifications be done almost instantly, but some of them may require a certain amount of time and memory, for example finding the shortest path between each stop in a route. To be effective, we must have a very fast computer and a high transmission rate. We have chosen to implement this system on a dedicated work station composed of a high resolution bitmap color screen with a MC68000 based microcomputer running a UNIX like operating system. This is a very expensive piece of equipment but we must realize that the operations we want to optimize are very costly ones where a small part of the expected savings can pay off the investment.

Interactive aspects. We have implemented a very simple interface consisting of menus and cursor positioning with a joystick, the keyboard is almost never needed. This allows direct manipulation of the objects (in our case the stops and the routes) which is recognized as one of the easiest way to interact with a system.

Another important aspect of the system is the possibility of undoing commands. If we realize that we made an error in specifying the routes or stops involved in a command or that a command did not give the expected improvements then, by touching one key, we can go back to the situation before that command. This feature is very important because in this way, one can freely experiment with the system without having to find the inverse of an erroneous command.

This tool is user-friendly, easy to use and cost effective. It has allowed users to evaluate and modify more than 150 routes in two days. We have focused here on the modification aspects of the editor but it can also be used to create new routes from scratch.

A Routing/Scheduling Editor

With this editor, we want to deal with problems of the "dial-a-ride" type: these problems have the following properties: when a vehicle is assigned to a client, it must pick him up in a certain time window and leave him at his destination in another time window. As all requests for services are taken in a central place, it is hoped that it will be possible to combine many of these so as to achieve a better use of the vehicles and a less expensive solution than the one that would dispatch a vehicle for each demand: in that case, a taxi service would solve the problem.

Savings are possible because of regular clients who have similar requests for consecutive days: e.g. a person who uses that service to go to work each day or to go to a regular appointment in an hospital. Sometimes organizations that set up such dial-a-ride services insist on receiving requests in advance. With all these requests, it is possible to create routes that will satisfy the time-window constraints and reuse the existing vehicles as much as possible. We were in fact involved in the development of a batch system to optimize routes in this way (see Roy et al. (1985)). The system can even deal with new requests that come on the fly and insert them in existing routes if possible, otherwise it asks for new vehicles.

To evaluate the results of that system and to modify them as needed, we have developed an editor adapted to that kind of route.

The previous section described a tool to evaluate routes where there was no notion of time; that is, the pickup or delivery time was not important. In the context of school busing, this was so because all students going at the same school were taken together and it was assumed that they all started classes at the same time and went to the same destination.

But, in our new context, this is not so and there are time-windows either on pickup or delivery. This means that the geometry or shape of the route is not anymore the only criterion for evaluating the routes. It might be acceptable for a route to loop on itself if the comeback is done much later. Similarly, we can only think of moving requests from one route to another if the pickup and delivery points are in the right vicinity of the existing route but also if they are in a time span compatible with the time-windows of the route. One has also to take into account the fact that inserting a new request in a route might change the pickup and delivery times of the requests following it. Similar problems might also arise from the removal of one stop from a route.

One characteristic of these routing/scheduling problems is the great amount of information necessary for a good evaluation; one has

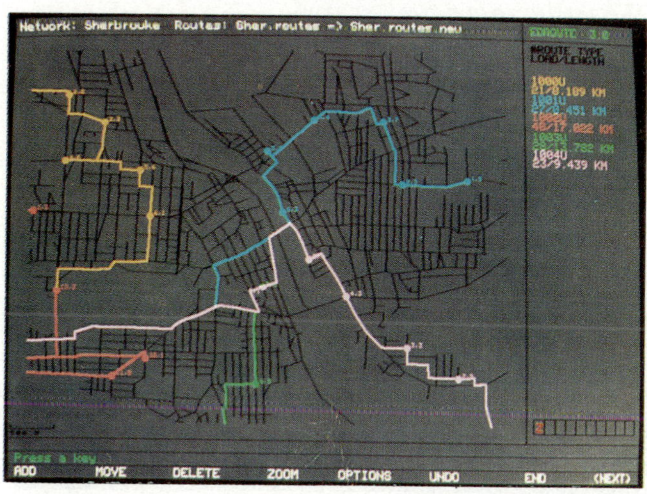

Figure 1. Interactive route editor. Five routes are shown over the network of Sherbrooke. At the bottom of the screen there is a menu of commands and on the right a zone giving information about the routes.

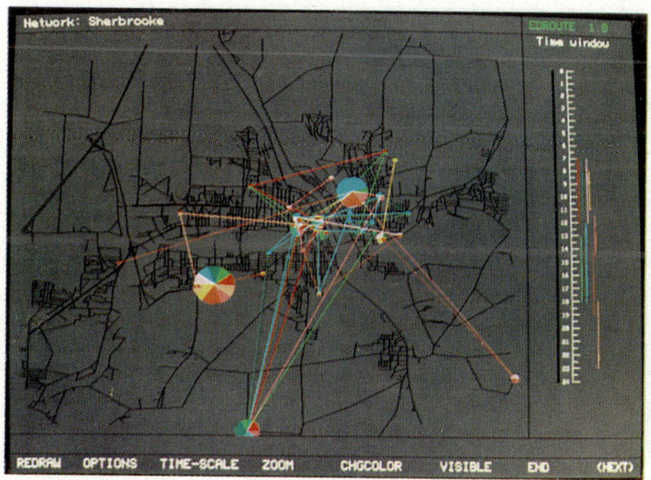

Figure 2. Routing/Scheduling editor. Nine routes spanning a whole day are shown: the black line on the left of the time scale indicates that the full day is displayed and the lines on the right side indicate for each route its time period.

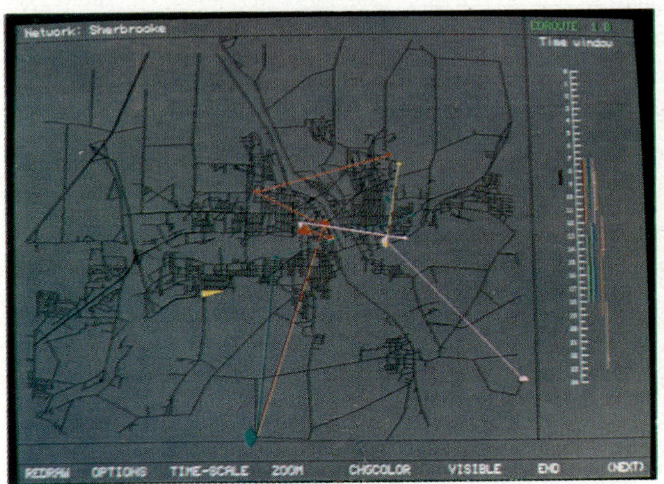

Figure 3. Routing/Scheduling editor. A window covering one hour, from 8:00 to 9:00 a.m., let us see the activity on the routes for that period.

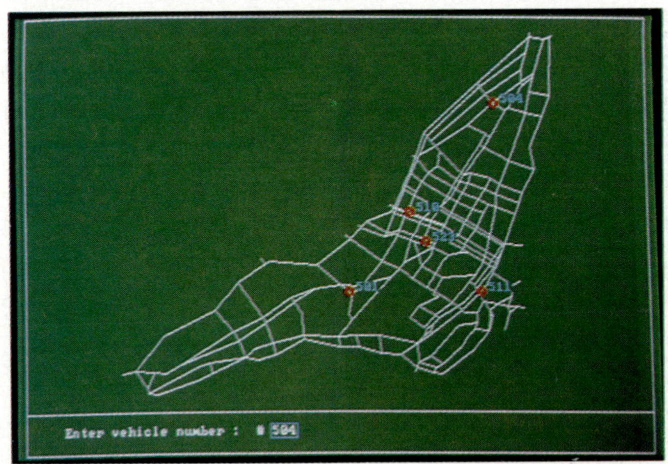

Figure 4. Dispatching program. Red dots with blue numbers show the position of the vehicles.

to see not only the geometry of the route but also the time they spend at each stop. Moreover, one must be able to identify where a given person is taking or leaving the bus: in other words, one must be able to locate quickly the two stops forming a request. Figure 2 shows nine routes spanning the whole day. On the right side, the lines with the same colors as the routes represent the time of the day when a vehicle is active and thick lines show when the vehicle has at least one passenger. As we can see, the drawing can be very loaded (even if we are still missing information like the time of stops) so our editor has a set of commands to restrict ourselves to a shorter time span. Figure 3 is showing what is happening between 8:00 and 9:00 for the same routes as Figure 2. With those commands, the user is able to determine the lower and upper bounds of the time span and to move it forward or backward. A useful way of working is to define a short time span and to move it by steps of one half-hour: so the activities of the whole day can be followed incrementally.

We spent a lot of time trying to find ways to display all that information in a meaningful manner and we must admit that the results are not entirely satisfactory. As soon as we have two or three complete routes or all the routes within a short time span, the display becomes too loaded: we are still searching actively to improve this aspect of the editor.

Our editor does not yet include modification commands but we plan to go as follows: if a stop seems to be in the wrong place, the user will point to it using the cursor and the program will then compute all acceptable positions in the other routes where that stop could go. The program will then flash those positions giving for each a value related to the inconvenience brought by the insertion of that stop in that route. From all this information, the user will be able to choose an acceptable spot within another route to move the "bad" stop.

We believe that in an interactive system, the user should always be able to override any automatic decision. One stop can be moved anywhere but the program will warn against any violation of constraint.

We have not yet tangible experience with that editor as it is still in development but we think that it is the "only" sensible way to go. Because there are so many things to take into account at the same time, we need an online automatic tool to check rapidly that the intended changes are feasible and do not bring too much inconvenience to the clients. At the present time, people who work in that area satisfy themselves with feasible solutions and try not to change them rapidly because they do not have the time or the patience to compute all the implications.

Real-Time Dispatching Problems

We have seen the use of interactive computer graphic techniques for the planning of routes. These techniques could also be very useful in the real-time control of operations like the dispatching of new orders that occur as the routes are being done.

We have studied the following scenario: vehicles are running all over town and as the day goes on, new orders arrive. An order consists of a pickup at one point and a delivery which has to be done within a short time span (e.g. an hour and a half) following the call. The drivers of the vehicle regularly report at the dispatcher's office and then receive a list of new jobs to do.

The short interval between the arrival of a request and its execution reduces the possibility of planning: decisions have to be taken in real-time. Before assigning a request to a vehicle, the dispatcher must take into account the position of the vehicles, their current destinations, the time to serve the request, the capacity of the vehicles and the effect of inserting a new order in a given route. As long as this assignment is not confirmed to the driver, the dispatcher can change his mind especially if new requests make it more profitable to do so but this approach is even more demanding because more information has to be kept in mind.

We are developing a system for keeping track of all the information about vehicles and orders and helping the dispatcher to find the best route in which to insert a new order.

In fact, this system will be a synthesis of the systems described in the previous sections: we need an interactive location tool to help find the position of the pickup and delivery points given their addresses, we use the graphical tool to display the road network and the last known position of the vehicles as shown by Figure 4; we also need a good algorithm to find the best vehicle for an order (we use a slight modification of the one used in our dial-a-ride system mentioned in the previous section), and finally we need a tool to modify routes and to try different scenarios.

This system helps a dispatcher to do a better job because it relieves him from having to keep track of all the details. Experience has shown that the dispatcher's job is very demanding and tiresome; the difference between a profitable business and a loss making one is usually the direct result of the dispatcher's job; a good tool to help him brings many savings (while keeping him in better health).

CONCLUSION

Interactive graphic techniques play a very useful role in the planning and implementation of routes and schedules because they allow

- time saving: routes are easily visualized and evaluated;
- improvements of the solutions: one can take into account constraints neglected by algorithms and modify the routes or try different strategies;
- interactions with people: dispatchers and planners are involved in the elaboration of the routes.

Now that powerful microcomputers can be equipped with high-resolution color graphic screen and fast hard disks at a relatively low cost compared to the savings expected, dedicated routing and dispatching systems can be developed. Such systems are based on a specialized interface that facilitates their use so that a dispatcher does not even have to learn how to use the microcomputer but only how to interact with the program itself. As the workstations are only starting to appear, more facilities can be expected in the years to come, algorithms that now have to be executed in batch mode on a mainframe computer will soon be feasible on those new machines. Still, due to the constraints and complexity of the problem, a human intervention will always be necessary to cope with the most unexpected situations, see Bloch (1977).

ACKNOWLEDGEMENTS

The authors wish to acknowledge Jean-Marc Rousseau and Jacques Ferland for their help and advices, all the people who work on the interactive graphic projects in particular Serge Roy, Pierre Cossette and Denis Derome, and Madeleine Achard for the word-processing of this manuscript. The development of the programs mentionned in this paper was supported for the most part by the Transport University Program of Transport Canada.

REFERENCES

Babin, A., Florian, M., James, L. and Spiess, H. (1982). Interactive graphic method for road and transit planning. Transportation Research Record 866, 1-9.

Bloch, A. (1977). Murphy's Law and Other Reasons why Things Go Wrong. Price/Stern/Sloan Publishers Inc., Los Angeles, 96 p.

Bodin, L., Golden, B., Assad, A. and Ball, M. (1983). The routing and scheduling of vehicles and crews: the state of the art. Comput. Ops. Rech. 9, 69-211.

Canadian Mining Journal (1975). Mount Wright, the computerized iron mine. Southam Business Publication, March 1975.

Cullen, F., Jarvis, J. and Ratliff, H. (1982). Interactive optimization in distribution analysis. ORSA/TIMS Joint National Meeting, San Diego, Oct. 1982.

Denardo, E. and Fox, B. (1979). Shortest route methods 1 reaching, pruning and buckets. Op. Res. 27, 161-186.

Dial, R., Glover, F., Karney, O. and Klingman, D. (1979). A computational analysis of alternative algorithms and labeling techniques for finding shortest path tree. Networks 9, 215-248.

Fisher, M., Greenfield, A. and Thompson, K. (1983). Real world experience with an interactive colour graphic interface for vehicle routing models. ORSA/TIMS Joint National Meeting, Orlando, Nov. 1983.

Roy, S., Rousseau, J.-M., Lapalme, G. and Ferland, J. (1985). Routing and scheduling for the transportation of the disabled persons: the algorithm. Centre de recherche sur les transports, Université de Montréal, pub. 412A, 21 p.

Tuan, P. (1984). The application of microcomputer and on-line graphical display technology to demand-responsive transportation services. Comput. Environ. Urban Systems 9, 1, 93-109.

EDITORS' NOTES

(1) Chapter 9 provides a review of the commonly encountered routeing and scheduling algorithms as applied to freight operations. Chapters 7 and 10 describe scheduling procedures used in the rail and bus industries respectively. They adopt a similar classification of the sub-problems involved, but different circumstances prevailing in the two industries are reflected in the solutions chosen.

(2) An evaluation of methods of matching applicants with one another in an organised car sharing scheme reached a similar conclusion; the ability of a human, particularly if he was familiar with the network, to identify sensible journeys and diversions could not be replicated adequately by a computer. The preferred solution was to have the computer plot the location of potential sharers on an overlay to a 1:50,000 scale map using a code to indicate special requirements (e.g. to reach the destination in such-and-such a time period or to share with non-smokers) and then to have a human select the best potential matches for each applicant. For further details, see Bonsall and Spencer (1981) "A comparison of manual and computerised matching procedures for organised car sharing", ITS, TN61, University of Leeds.

(3) See Chapter 14 for further details of the technology that might be involved in such systems.

Information Technology Applications in Transport, pp. 257–286
P. Bonsall, M. Bell (Editors)

Chapter 12

COMMUNICATIONS TECHNOLOGY IN THE TRAVEL INDUSTRY

Richard Pope
Fir Tree Consultancy, UK

THE ADOPTION OF TECHNOLOGY IN THE TRAVEL INDUSTRY.

Introduction.

The travel industry is well known for its use of information technology but, although the public perception may be of continued and rapid growth since the early 1960´s, most progress has in fact been concentrated within the last five years prior to which there were, by comparison, some 15 years stagnation! Recent progress has been following an exponential curve which raises some interesting questions as to whether the industry can continue to absorb the developments at the rate they occur.

The Needs of Airlines.

The Inadequacy of Manual Systems. The increase in business and pleasure traffic and the increase in route permutations created severe problems for airlines particularly in the United States; passenger bookings became less predictable, more often changed at the last minute, cancelled or doomed to be classed as ´no-shows´. Commercially available computers were the saviour of over-stretched manual ´reservation rooms´ and for the computer manufacturers themselves, this was a prestige application.

On-line Systems. The need to hold reservation data centrally, but accessed remotely by airline offices all over the world, produced bigger and bigger systems catering for more and more data and quicker and quicker response times. The airlines now operate some of the largest and most powerful commercially available computer systems; an operation of 20 - 30,000 terminals on-line is common.

Marketing Tool. Competition between carriers forced further advances as a reservation system became a marketing tool as well as an operational tool. Airlines today have extensive facilities over and above reservations. They are able to print boarding cards to be issued with the ticket (thus saving the ´executive´ time by not needing to check-in before a flight). They are able offer seat selection at the time of booking the flight. They are able to give weather, theatre, resort and in-flight entertainment details. All of which act as a marketing tool by encouraging the passenger or agent to book with the carrier whose system offers the widest range of facilities.

Perhaps the most important marketing tool within an airline (1) reservation system, though, is its ability to quote fares. Obviously fare calculation facilities significantly help airline staff by avoiding lengthy manual effort but in marketing, the ability to introduce or change any fare, limit its availability of or alter qualification requirements at a moment´s notice provides the airlines with their most important marketing tool. This facility is, of course, particularly important in the deregulated US environment. Fares and fare structures are now being changed to react to other carrier´s pricing policy to such an extent that changes can reach 5000 in a day and 20,000 in a week! - The astute viewer of television advertisements can spot voice dubbing where fare changes cannot be reflected in time without both expensive re-filming of the commercial and interruptions to the campaign schedules!

National Differences. The sophistication of airline systems varies from country to country. Competitive pressures within the USA, even before deregulation, stimulated the most rapid and extensive advances; facilities found within US airline systems are even more extensive than those already described.

Elsewhere most countries have a national carrier and competition is less fierce. Route allocations, bilateral fare pricing, shared cost / revenue arrangements between ´competing´ carriers (known as ´pool partnering´), etc., reduce the pressure for more and more facilities to be added to an airline system.

However the nature of international travel requires that the most basic reservation system has to take into account the requirements of a passenger travelling world-wide. The passenger increasingly demands the lowest fare even if it means altering his itinerary or buying tickets in other countries. Airline systems the world over are thus being forced to increase their sophistication even when the airline itself will probably not benefit at all.

The Needs of Travel Agents.

The Basic Requirement. Travel agents have always been packagers of travel components from many suppliers. Their sphere of operations can cover the business traveller, the leisure market, the group movement, the incentive market, the high cost product, the low cost product. In fact an agent can be asked to provide any travel product conceivably available and so has to have full knowledge of each and every product.

Against this background it is perhaps surprising that travel agents installed so little information technology prior to 1980 but it was not until the advent of airline systems that the travel agent had an opportunity to do so. Even today the penetration in most countries other than the USA and UK is limited to those agents who primarily handle business travel.

On-line Systems. During the mid-1970´s airlines had developed their reservation systems to such an extent that competitive pressures forced travel agents to install terminals linked to the airline systems.

(It should be noted that terminals are supplied by the airline as a package. The agent does not have any choice as to what type of terminal he has and more importantly, it cannot be used to connect to other systems however much investment in communication equipment the agent is prepared to make.)

The advantages of a sophisticated productivity tool were only slightly offset by the fact that many airline systems had, and some still have, an in built bias to favour the airline whose system it was. Depending upon which country one is in, this was (is) either treated quietly or used to marketing advantage.

A more serious drawback is that of cost. Airline systems have by their nature to be central on-line systems. These in turn necessitate dedicated-line communication between the terminal in the travel agent´s office and the central computer of the airline. The USA has always had low cost dedicated-line communications and therefore the spread of travel agent automation was rapid. But where dedicated-line communication costs are high (ie. everywhere else) the traditionally low profit margins prohibit travel agents installing such equipment. The slower, but significantly cheaper, telephone has sufficed - if not ruled!.

Marketing Tool. As with airlines, the installation of reservation terminals in travel agents did have marketing advantages. Primarily it attracted business and differentiated agents selling the same product.

During the late 1970´s and early 1980´s, the major travel agent chains in Europe installed airline terminals in most of their offices primarily for this marketing advantage. As a productivity tool it was excellent too, but only for flights (products) where that airline´s terminal could be used. Bias and poor connections to other, probably competitive, airline systems prevented better use.

Multi-access. As has already been explained, airline systems in general were quite restrictive and whilst fine for an airline, they were not ideal for travel agents who require no restrictions. In parallel with the expansion of airline terminals into agents, some airlines, governments, consortiums and travel agent associations saw an opportunity to overcome those restrictions and develop on-line systems that could switch between different carrier and/or other supplier systems.

These developments occurred in many countries. For example Travicom in the UK, START in Germany, SMART in Scandinavia, TIAS in Australia, MAARS+ and Tymeshare in USA, TIMAS in Ireland and Esterel in France.

All were/are based on dedicated terminals accessing a central computer which onward connected into the appropriate reservation system (see figure 1).

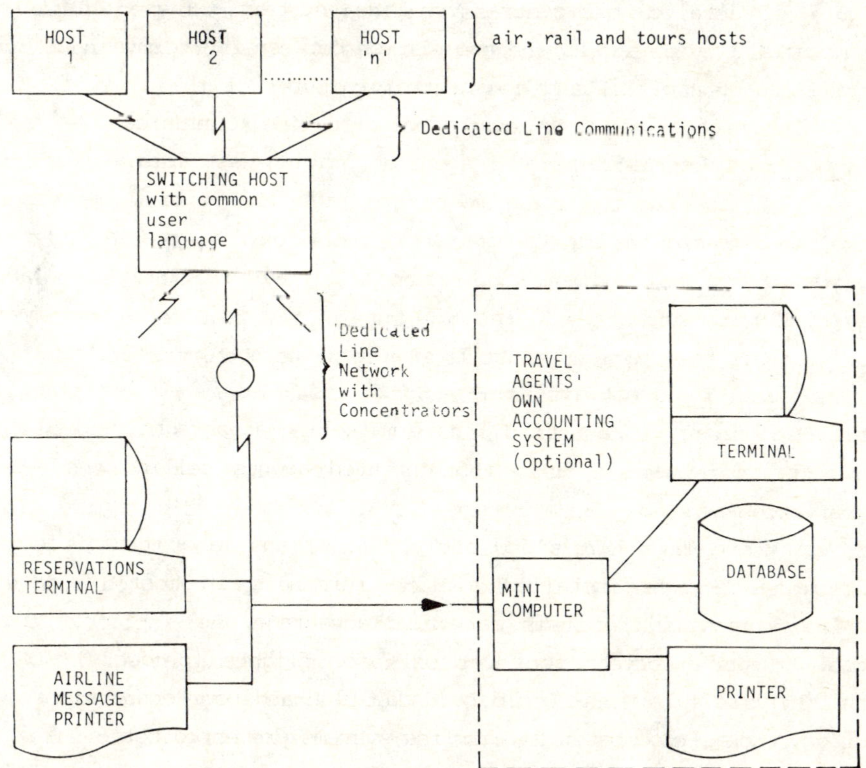

FIGURE 1 : TYPICAL MULTI-ACCESS, RESERVATION ONLY, SYSTEM

Depending upon the system, the central computer may have been one of the airline hosts and in many cases acted as a (user) language translator, a ticketing and document printing system and, in conjunction with dedicated mini-computers attached to the agent´s terminal, a travel agent´s internal accounting system.

Generally speaking, the USA multi-access systems failed due to competitive pressures from the major carrier´s own systems which, in a variety of ways, provided their own multi-access capabilities and travel agent accounting. Elsewhere, the systems were more successful despite limitations on direct access, particularly to non-national carriers. In Europe (except UK) multi-access is usually available for a combination of rail, tours and ferry reservations.

Even with the advent of these multi-access systems, there are a number of detailed restrictions and drawbacks; most particularly their inability to access the ever increasing number of new industry (particularly non-air) suppliers economically.

The Needs of Hotel Chains.

Like most companies that consider automation, hotels installed systems for their own internal benefits. They invariably handled administration aspects and did not immediately require systems to handle their reservations apart from within the hotel itself.

Traditionally hotel reservations have been made by telephone, telex and post; computerised reservations have been slow to catch on. The emergence of major hotel chains created the need for reservation systems and because of their association with air travel, there was a natural link with them.

In many instances an airline which owns or is connected with a chain of hotels will itself hold details of hotel availability; the hotels supplying details via regular tape or telex updates. The USA has always had a large number of hotel chains and coverage, as already explained, was similar to the airline network. For the rest of the world, there are a significantly greater number of independent hotels. Many of these have formed consortiums to operate a reservation service for members. But few such systems have links with airline or other such systems. In addition, these reservation services often only act as clearing houses for surplus accommodation. Accurate availability is restricted to the individual hotel´s internal system.

The Needs of Car Hire Companies.

The major chains needed computer systems to keep track of their cars which could, of course, be anywhere at any given point in time. If hire usage was to be maximised, a reservation system had to be developed.

The difference between the USA and the rest of the world is quite marked. The big four within the USA - Hertz, Avis, National and Budget - have national coverage and for some time have had links with the airline systems. Other car hire companies, operating within only one or two States, now also have links with airline systems but some of the links are quite unsophisticated. Outside of the USA, the big four (National is linked with other chains outside the USA. - eg. Europcar) are still represented in airline systems albeit not all by direct access links. Often reservations are made under free sale agreements. (ie. Book the car hire and report it as sold unless there is a short lead time when a confirmation has to be waited for.)

The major car hire companies are only just beginning to market independently of airline related campaigns. The increasing popularity of leisure market products with a hire-car component (eg. fly-drive tours and villa holidays) is a particularly significant development.

Car hire companies are also directing themselves at the travel trade as a whole in a bid to expand away from the predominately business travel product. With few exceptions, their reservation technology, however, is not geared up to direct access other than via the airline systems and since many airlines have associate or preferred car hire companies, this can act to the disadvantage of the consumer and car hire company alike - surely a market imperfection which will not survive in an increasingly competitive environment.

The Needs of Ferry Companies.

Ferry companies have not, in the past, had the need to install reservation systems or, if they have, it has been a low priority. However, within Europe, the late 1970´s and early 1980´s have seen a big expansion of ferry based holidays tied in with other products such as camping, villas, hotels and even day trips. These new ventures demand the type of planning and control made possible through reservation systems. Information technology is accordingly beginning to play a priority role in this leisure market. The business market, which is usually the first to benefit from technological investment, is less advanced; it only now being attacked in areas such as Scandinavia where ferries are as essential as airlines.

The Needs of Tour Operators.

Large independent tour operators have needed computer systems for many years just to administer the many components of any tour brochure. When brochures are published, there can be no way of knowing which hotels will be sold with which seats on which aircraft. Frequently there will be a flight seat available but not hotel space and vice versa.

National carrier based tour reservations are often held on the same computer system as scheduled seats since they use the same flights. Hotel availability, too, can be found on the same system as this component is often contracted on an allocation basis. National carriers operating significant charter flight based tours normally have a sister reservation system handling this traffic. In this latter case, national carriers operate more like independent tour operators.

Tour operators use their systems to identify and manage trends allowing them to release hotel space for lack of demand, add flights because of demand, or consolidate two flights from different departure airports into one flight to one destination. In order to allow operators to take any action to reduce costs, their systems have to be very comprehensive.

Late Booking Trends. Because tour operators are permutating many component allocations in any one brochure, a six to eight week lead time for finalising all bookings prior to departure has been necessary so that unused allocation can be released without penalty from the supplier and so that tour documentation can be processed and distributed to clients and the resort.

In some countries and particularly the UK, this lead time has been eroded with last minute bookings becoming an established option. (Massive over-capacity by UK tour operators in the early 1980´s combined with television programmes suggesting reduced prices for those making last minute bookings, were the causes of this development.)

Computer systems geared for the traditional pattern are unable to cope with late availability transactions. The problem for the tour operator was how to inform the UK´s 6000+ travel agents of the actual availability when existing procedures were only designed to issue (broadcast) telex messages of such availability. As late availability had not previously been a significant factor, even the telex production was often manual!

Videotex - the UK Experience. Coincidently with the growth in late booking, the UK videotex service, Prestel, had been established and had attracted a significant number of travel information providers. The Prestel system did not (at that time) have interactive facilities but did provide excellent direct look-up of data pages at low cost.

Sealink, the UK ferry operator, had embarked on a pilot trial using Prestel to hold availability information as a more up-to-date substitute for issuing printed availability charts. Part of this pilot scheme was to subsidise videotex terminals to agents - an offer most accepted. The tour operators, therefore, had an existing installed base of technology with which they could distribute late availability information. The result was an instant success and videotex became a permanent feature of travel technology. In peak season, some 1.5 - 2 million accesses for late availability can be registered per month.

The wide-spread use of videotex terminals and their ability to be connected to mini-computers front-ending a ´normal´ mainframe computer provided a low cost solution for both tour operators and agents to make direct access tour reservations a reality for the first time. Within 2 to 3 years the number of videotex systems available within the UK for reservations has exceeded 30. Currently they are accessed through a mixture of direct dial, dedicated networks, Prestel gateways (ie. access to other systems via Prestel), private networks and network operators. This mixture of access points has not deterred the agents: the major suppliers report that some 90% (4 million holidays) of bookings are made this way instead of via the traditional telephone.

Mainly because of this adaptability, videotex has already gained a measurable acceptance throughout Europe and Australia as the travel industry´s normal communication method. Whilst all videotex systems are not completely standard, their concept is. Their ability to link into and use packet switched networks has given any videotex system the capability to switch and communicate on an international level with an almost infinite number of hosts. This ability can clearly be seen as the essential feature of a travel industry tool which had been lacking in previous systems.

Another role for videotex was found by the newly expanding numbers of small and specialist operators offering a wide range of products from weekend breaks to windsurfing, from villa programmes to skiing.

To both cater for the demand and to compete on equal (videotex) exposure terms, these operators were faced with having to install reservation systems. The sophistication now being reached by the desk top personal computer (PC) allowed many to use these machines as reservation systems. By also installing auto-answer modems, the operator was then able to allow direct access for reservations by agents. Videotex had thus created a significantly lower cost entry point than had previously been possible.

Once the trend had started, suppliers from all sectors such as coach, charter flight specialists, air schedule guides, tourist boards, etc. started to use videotex technology, whether interactive or not, to promote and sell their products. For example the English Tourist Board developed a reservation system to hold availability on hotels and guest houses that otherwise would not have had booking facilities and were therefore unlikely to have been sold through travel agents.

THE TRAVEL INDUSTRY IN 1986.

The Current Position.

The Technology Adopted. Central on-line systems developed to solve the needs of the airline industry now dominate the industry. Dedicated computer terminals providing comprehensive facilities can be found in almost any travel location around the world.

Extensive computer and communication power provides the airline with the means to service the traveller with amazing efficiency. Bad weather and missed flights mean whole itineraries can (instantly) be recompiled. Accommodation and special requests can be arranged at even the smallest hotel, all with response times of two to three seconds.

The leisure and non-air suppliers do not have extensive networks but within their own sphere, their systems are of comparable sophistication. The linking of these systems to videotex to create the networks has created pressure on the airlines to make their own systems accessible from terminals other than those supplied by the airline.

The Travel Product. Travel itself has undergone a change; there are now many more variants of the same product, many more suppliers, travellers and interested parties such as government tourist offices promoting their own countries. There are also many more packagers of travel products such as companies organising conferences or incentive travel. This diversification and the overall growth in size of the industry has put increasing demands on technology to solve the resulting problems.

The Technical Requirements.

Different Requirements of Different Suppliers. The on-line systems created suit the major suppliers well with one exception: interconnection between systems is expensive and so is unlikely to be adopted unless it brings increased revenue (as in the case of the links developed by airlines, hotels and car hire companies). Market forces cause the major travel suppliers to interconnect with as many complimentary suppliers as possible in the most effective manner.

The options for other travel suppliers are more difficult. The cost of establishing systems in a similar way to airlines is obviously prohibitive. The ´cost´ of choosing one of the existing airline systems for distribution of their products is fundamentally a question of what restrictions will result; which principal does a company choose and what are the connection terms? A company must also consider its own internal requirements first and this may conflict with connections to reservation systems. Will a system that is good for administration be suitable for other tasks?

Requirements of Travel Agents. Travel agents require maximum flexibility for reservations with minimum response times. But travel agents have their own computer needs too, consider for a moment their other technological requirements:

Telex facilities are still a main source of communication used for a wide range of hotel reservations and confirmation, special requests and frequently used as an information medium for tour operators whether it be for late availability or a change in a client´s itinerary.

Accounting systems are another essential requirement. So far technology has concentrated on the reservation aspects, but keeping track of client purchases, foreign exchange transactions, commission owed, deposit payments, etc., has to be a high priority.

Information databases in their own right are becoming increasingly important. Both as a way of keeping ahead of the competition and of quickly refering to large quantities of information. Travel agents are developing their own databases for air fares, tour availability, promotional products, etc..

Video as marketing, promotional and training tools are rapidly gaining ground for travel agents and tour operators alike.

Expenditure on all these and other items are in addition to airline and videotex terminals. The answer to most of these problems must be the packaging of different products together to give a greater flexibility without at the same time the need for a multitude of equipment.

Developments Within the Industry.

START - West Germany. One of the first attempts to tackle the need to access different reservation systems and perform other functions, such as document printing, was a government supported consortium in West Germany called START. Formed in the early 1970´s, the aim was to provide multi-access reservation facilities and ticket and travel documentation for air, tours and rail; and accounting facilities for agents.

The system (shown in figure 2) gave travel agents on-line access via dedicated-lines to a central computer in Frankfurt. The central computer could then switch into the Lufthansa systems for air reservations and air ticketing, into the Deutch Bundesbahn for international and domestic rail reservations, into the TUI reservation computers (itself a consortium for several major tour operators) for tour reservation and booking documentation. The remaining functions are carried out by START computers with the exception of (travel agent) accounting, which was handled by a link with a bureau accounting service operated by DER.

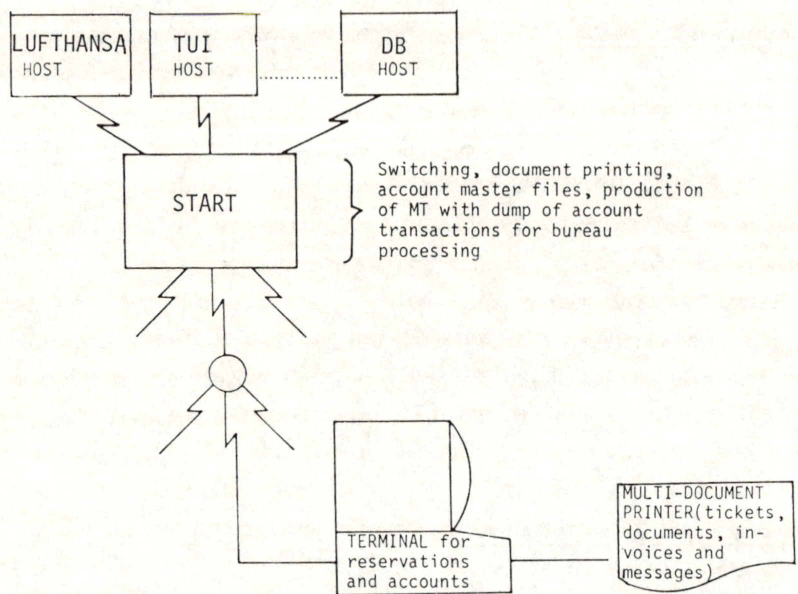

FIGURE 2 : THE START SYSTEM

Conceptually the design was ideal because it tied together the major travel suppliers and overcame protocol standards, etc.. The service, whilst delayed, has operated well from 1980 but still suffers from several drawbacks:

* The equipment is dedicated for START purposes and the facilities cannot be tailored for any one agent.

* Connection to START is governed by consortium partners and has, at least in the past, hindered connection by competitive systems. In particular other airlines were at first not allowed to connect and then, when they were, the connection was on the basis that their flight schedules were only displayed after the agent specifically requested further flights other than the Lufthansa flights shown on the initial display.

* Connection of additional reservation systems to the system involves expensive programming within the START computers.

* The cost to travel agents is high although tariff changes were made to reduce the impact on those agents not requiring all of the facilities.

In recognition of potentially competitive pressure from a national videotex network, the START system now has videotex access through to the supplier systems.

TIMAS - Ireland. The TIMAS system in Ireland is a similar attempt by consortium partners to establish a multi-access system for the travel trade.

Originally the only system available to agents was the Aer Lingus system that was typical of the systems available during the late 1970's. The limitation of this single-access system was challenged by international airlines who established a competitive multi-access system. Pressure from agents wanting (needing) Aer Lingus system capabilities and financial difficulties of the multi-access system without Aer Lingus resulted in a consortium being established to operate a new service combining the two.

The consortium is owned by Aer Lingus, the Irish Travel Agents Association and a computer company called Videcom. TIMAS is configured as in figure 1 but without a linked accounting system. The hosts are the airlines: Aer Lingus, British Airways, Dan Air, TWA, Northwest Orient, Air France and Lufthansa, the tour operators: JWT and Aer Lingus holidays, and the ferry companies: B & I line and Sealink.

The system is based on a system developed by Videcom (and implemented in several other countries) consisting of 100+ terminals, connected via five satellite concentrators spread around Ireland, and linked through the Videcom communications computer into the different host computers.

The TIMAS computers switch between hosts by user selection and interpret the common user input language into the format that the host can understand. However, to avoid complexity and a degradation of response times, responses are not converted. Neither does the Videcom computer perform other functions such as accounting or information retrieval from a reference database.

Travicom and Skytrack - UK. The UK has a wide spectrum of travel products (and thus reservation systems) and a competitive airline industry with many international carriers. Against this background British Airways, British Caledonian and others cooperated to form a system called Travicom based, like TIMAS, on the Videcom system and with a configuration similar to figure 1. The original system first became available in 1978 but there have been important developments since then.

Despite having connections to many airlines, the original system was regarded as expensive and inflexible. Also prior to 1981, the licensing restrictions of the PTT (post, telephone and telecommunications authority) prevented tour operators and other non-air hosts being on the system. When these restrictions were lifted, the opportunity was taken to initiate a project to redesign the system around a packet switched network (PSS) and to provide ´soft´ connections to hosts (thus avoiding the separate circuit boards required for ´hard´ connections). The aim was to create a full multi-access system for the industry. In the event the project failed due to high development costs, poor (pilot) response times and sheer complexity. In hindsight, this type of central system would not have been suitable for the UK travel industry anyway.

At about the same time the tour operators began using videotex technology. The problem was then how to provide an agent with the ability to access traditional airline systems and videotex. The solution adopted by Travicom was to use a dual standard terminal that could simply be switched between Travicom (via a dedicated-line) and videotex services (via an auto-dial modem and ordinary Public Switched Telephone (PSTN) lines). The inconvenience inherent in this procedure coupled with the fact that the terminal is not a colour one, has meant that they are predominately only for the Travicom connection and thus almost exclusively found in business travel departments. Their leisure travel departments use dedicated videotex terminals.

There obviously remained a sizable number of agents requiring a system combining leisure products with the lower volume air business as well as business agents requiring a system integrated with an internal information database and accounting system.

At the same time as tour operators were utilising the UK videotex system (Prestel) for late availability of tours, Prestel were implementing the gateway technology developed by West Germany - which was itself an extension to the original Prestel standard. Gateways allow the videotex user to carry out interactive processing and so Prestel conceived the idea of connecting a multi-access (airline) reservation system to Prestel as a gateway. They chose the Videcom system and with Videcom developed a service called Skytrack (see figure 3) to compete with Travicom but using videotex technology.

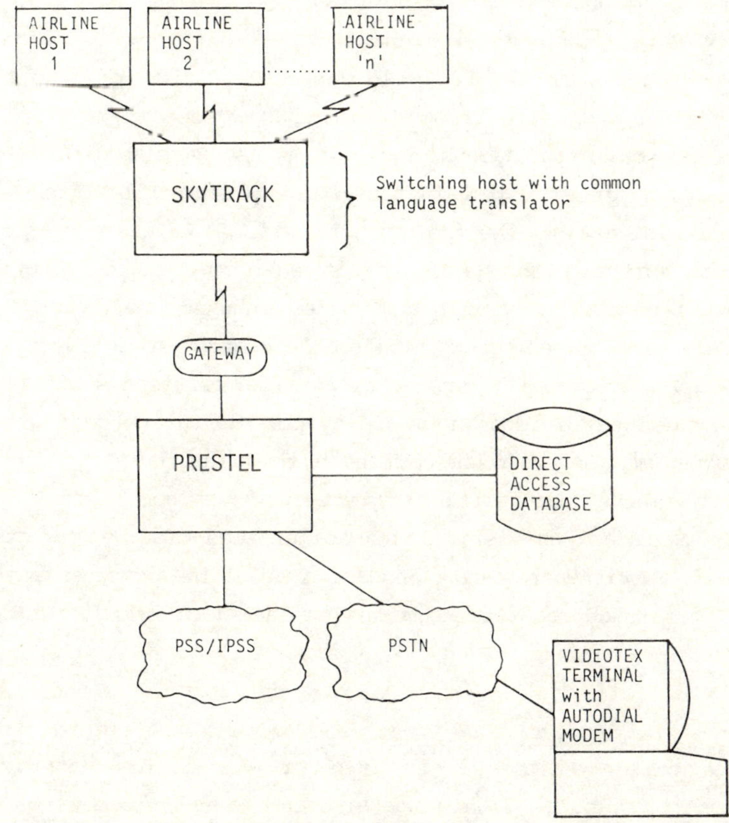

FIGURE 3 : THE SKYTRACK SYSTEM

Despite having no more than eight subscribing airlines, it was possible to use Skytrack to make bookings on all airlines because several of the subscribing airlines agreed to waive their ´host checks´*. For the first time ever it was therefore possible to use one terminal and service for all types of travel booking.

The obvious competitive pressure from Skytrack prompted Travicom to develop a facility to allow videotex access to their existing system. The service was called Travicom Traveller and had its own UK network in parallel to the existing Travicom (now named Travicom Executive) service network.

The Skytrack service did not succeed commercially due to the dominance of Travicom (with by then some 1,300 terminals installed) and the general assumption by agents that even the lower cost Skytrack service should have response times equal to Travicom (Executive); which of course it could not due to the different technology. Thus business travel would not change from Travicom to Skytrack and non-business agents did not require scheduled air reservations in sufficient volume to justify its commercial success.

Following an initiative by the Association of British Travel Agents and through commercial pressure, the Prestel Skytrack and Travicom Traveller services subsequently combined and the majority of Travicom facilities are now available through Prestel. The resulting system is shown in figure 4. One of the criticisms of the existing system is that it does not have a ´reverse gateway´ and so Prestel services are not available through the Travicom Executive system (except via the dual function terminals).

As has been mentioned earlier, travel agents need other forms of automation such as ticket printing and accounting. The complexities of air fares, rules and regulations mean that ticket printing is not an easy exercise.

* (´Host checks´ are limitations imposed by airlines to ensure that at least one segment of the flight itinerary is on the carrier through which one is booking. Apart from ´using´ one carrier´s system for convenience, the limitation also relates to whose ticket one uses for the journey. As the agent must pay the airline whose ticket is used, irrespective of whose sector the flights are on, that airline then has to reimburse the other carrier(s) for their portion.)

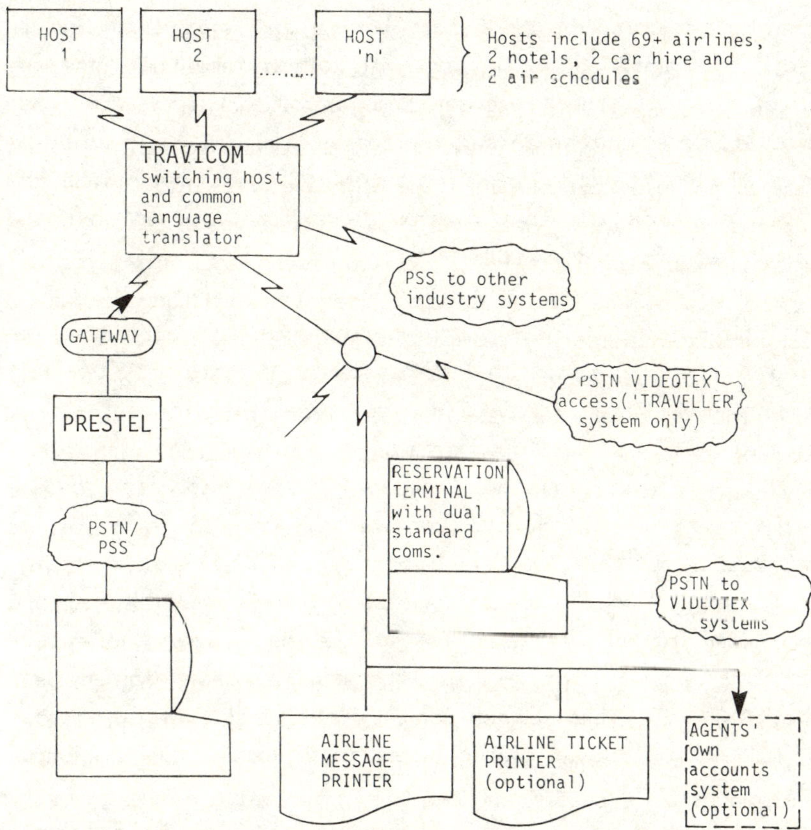

FIGURE 4 : THE TRAVICOM SYSTEM

The options to provide such a service are limited to either the host system or a linked computer in the travel agent´s office. With a multi-access system like Travicom where each host is entirely separate, it is impossible for each of the 60+ airlines to provide one integrated ticket and accounting system. For Travicom to do this, one would be using a computer designed for communications switching to process a significant commercial application. The START system performs some of these accounting and printing processes and it initially suffered from severe capacity problems that were only solved by a considerable increase of computer power.

The option chosen by Travicom was to have a linked computer within the travel agent´s location. Conceptually, the host computer thinks that the linked computer is a dumb device and when ticketing is required, the agents add ticketing information into the ´remarks´ section of the PNR (the Passenger Name Record which is the base record of all reservations in any airline system) and the host dumps a copy of the PNR to the linked computer. This in turn prints the ticket and accounts for the sale to the client. Perhaps not the ideal situation but it does work!

USA Airline Systems. The USA environment has already been described as one of large on-line systems. The few multi-access developments have largely failed due to competitive pressures from the major existing US carriers systems.

Instead the main airline systems have gradually become multi-access systems in their own right. The largest system, American Airlines SABRE, has evolved into a multi-access system using a variety of connections. Some are via the central system and some are via ´by-pass´ mode allowing the user agent to select another host carrier, but without the benefit of a common user language. This also has limitations as the facilities available within SABRE are not available when in by-pass mode to other airlines.

Eastern Airlines with their SystemOne service has recently developed a multi-access system connecting into some 22 airlines. A cross between Travicom and SABRE, the switching is handled within the central system. Unlike Travicom and SABRE by-pass, any component flight sector booked on another host airline system is automatically copied into one PNR within SystemOne. Thus allowing a full range of ticketing and accounting facilities to be automatic as well. A significant advantage.

Videotex Networks.

The dominance of airline systems within the travel industry can easily be understood. However the development in a very short space of time of videotex networks must not be overlooked. The adoption of videotex within most countries where it has been introduced has been so rapid that no country can now afford to ignore it.

The very low cost terminals, the world-wide PSTN network already existing in each country for communication, the availability of cheap look-up databases (the PTT services like Prestel), the ability to connect front-end processors to existing mainframe computers very easily, have led to the existence of many, perhaps primitive by some standards, reservation services. Once services are in existence, the task then becomes how to improve on communications.

Today´s generation of videotex is based on PSS networks with videotex access points (VPADS) handling the user interface and the host databases. An obvious improvement for the existing services would be to remove the temporary front ends and the dedicated private networks and replace everything with quality PSS networks.

PSS with X25 protocol is now a worldwide standard, thus almost all host systems can be linked to PSS networks. Additionally the same terminals that were designed for the original public videotex services are still valid for today´s PSS videotex networks. An important consideration for an industry that operates on 9 - 10% profit margins and still has to invest heavily in information technology.

PSS networks are being established in many countries around the world including the USA (although there the dependence is still on dedicated-line networks). Videotex projects using these principles are already under way. France with their Minitel service is probably the most advanced in terms of terminal population as it approaches 1 million. Elsewhere in Europe, most countries are establishing such services. The UK, whose PTT environment is the most deregulated, has three such networks for the travel trade alone.

The Role of the Travel Agent Associations.

The Association of British Travel Agents (ABTA). In an industry where travel agents have relied on information technology not specifically developed for them, the automation penetration is remarkably high. In a few instances, ABTA included, trade associations have funded technology investments. Because of the vast sums required to develop anything like an airline system, these investments have almost exclusively been devoted to agent´s accounting applications.

With the existence of the UK Prestel service and the coincidence of the late booking trends described earlier, ABTA established an operational service. This acted as a clearing house for late availability from its tour operator members to its retail agent members.

The success was dramatic. From that point on, ABTA acted as a catalyst for a wide variety of developments within the UK travel trade. The culmination in its involvement was in assisting Prestel to establish a service specifically for the travel trade. Facilities include a quick reference index geared to agent requirements, access restricted to members only, an ABTA news and information database for members and preferential charging structures. The Travicom - Prestel link described earlier is also part of the service and other developments relating to training are planned.

The Australian Federation of Travel Agents (AFTA). In a similar way as ABTA has done, AFTA has encouraged the establishing of a videotex network for the Australian travel trade. Established in 1984, the service had expanded to some 700 terminals in operation by July 1985.

As elsewhere, the Australian travel trade originally had single-access airline systems up until the mid to late 1970´s. In response to pressure from travel agents and from trends in other countries, a multi-access system was developed and called TIAS.

Again due to end user costs and the emergence of videotex technology, there was a need for a lower cost option. Led by AFTA, the AFTEL project was evolved to support, not only agents requiring a high volume business system such as TIAS, but also a lower cost videotex system. The resulting system (shown in figure 5) has both gateway and reverse gateway facilities between the two and so a fully comprehensive service is available via either.

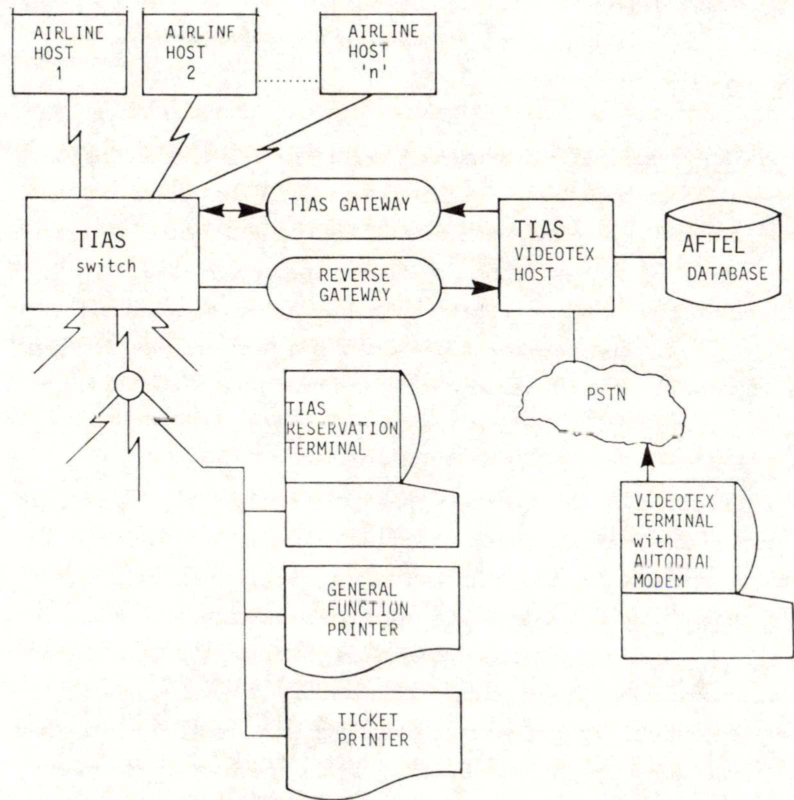

FIGURE 5 : THE TIAS/AFTEL SYSTEM

Other Travel Agent Associations. There are a number of similar projects progressing in various stages to those detailed above. They include New Zealand, France and Scandinavia.

Within the US there was little need for the trade association to initiate similar projects as the existing airline systems covered almost all reservation needs. Most accounting needs were covered through either the airline host printing invoices and tickets or by full accounting being performed through mini-computer systems linked to and marketed by the major airlines. (Accounting itself is less complex in the US due to the high percentage of credit card transactions and most international travel suppliers settling commission to agents in dollars.)

TECHNOLOGICAL DEVELOPMENTS.

Communication Networks.

Established Systems. The drawbacks of large on-line systems have already been documented and as time goes on, the problems will inevitably multiply as end users need more and more dedicated terminals on the desk in front of them. The attempts to-date in interconnecting these systems have brought about mixed benefits. Complexities, costs and the political question such as who develops and runs them will probably ensure that central switching computers will not be the answer.

Videotex. Every type of business has its entry level threshold when it comes to installing technology. Videotex provided the low cost, unsophisticated entry point that many companies on both the retail and supply sides needed. Videotex also demonstrated that low cost services could provide worthwhile services that many established systems had dismissed.

The development of videotex services has also established that technology can advance and compliment other developments, namely PSS. Through this association, videotex also allows the flexibility of connection that will enable a range of terminal types access to virtually any network world-wide.

PSS Networks. PSS has become a world standard for a variety of reasons: First, it has overcome the very high costs associated with dedicated-line communication in most parts of the world. Dedicated lines can only be efficient if used for a high proportion of the time.

Second, it has enabled network interconnection to be standardised (or computer to computer interconnection if Open System Interconnection - OSI - standards are included).

Third, by creating the role of a networker who himself can adopt any combination of communication technology, be it satellites, fibre optics or conventional data lines, flexibility and reduced costs are achieved.

Finally, because national PSS networks are compatible, international communication is achievable.

Desk Top Technology. The traditional dumb terminal sitting on the users desk is fast disappearing as the only suitable device. Personal Computer (PC) technology now gives the terminal device processing intelligence that can be used for both intelligent communicating and/or local processing. (2)

PC´s acting as simple terminals are now common in many environments where local intelligence simplifies log-on and search selection criteria of mainframe databases. PC´s can integrate reservations data with data for local use such as accounting.

These PC´s can in turn be part of a local area network within the office in which they are situated. Thus, under such software control as windowing, the PC can interact with both internal and external systems. The user applications are becoming at the same time easier and more sophisticated as the local intelligence hides complex application and technical procedures. The introduction of knowledge based systems will be another significant factor in making systems easier to use. (3)

It is perhaps not necessary to add that in turn these Local Area network´s (LAN´s) can be connected to in-house networks that are in turn connected to the external networks. This route provides a natural path to full Value Added Networks (VAN´s).

VAN´s. The product spectrum of the travel industry as a whole creates its own problems merely because it is so vast. One of the original principles of travel agents is that any agent can sell any travel related product or package of products. That fact demands the most sophisticated level of communication possible.

VAN´s, by their very nature are ideal for the travel industry. The ability to use a computer sitting on the network to assist in identifying, selecting, gathering and collating travel products can only be of the most significant benefit to the trade. Again, in conjunction with PSS network switching capabilities, the permutations become endless.

VAN´s will soon allow any agent via a videotex terminal, a PC, a LAN or other network to access another computer on the VAN network to ´assist´ with the travel enquiry. When that feature becomes a reality, information technology within the travel trade will have a new meaning.

Databases.

Reference Databases. Airline systems, of course, have extensive databases containing schedule guides for competitive carriers as well as their own flights. With many hundreds of airlines around the world, the gathering and collation of this data is a sizable task in itself.

ABC and OAG have for many years been compiling printed air schedule guides from data stored in their own databases for use by the travel trade and the business executive. Indeed airlines themselves use these databases as a common source to obtain the schedules of other carriers. Airlines have become sellers and buyers of the same data!

Since airlines cannot afford to store all air schedule permutations within their own systems, ABC and OAG have made their databases available direct. Initially available on videotex systems in the UK, France, Australia and the USA, more sophisticated links have now been developed with multi-access systems like Travicom. Both ABC and OAG databases include extensive fare details and are aguably the only way fare changes can be accurately tracked.

The travel trade has so far resisted the introduction of such reference databases. Whilst understanding such comments as ´you can put a pencil in a book to mark your place, but not on a vdu!´; we must conclude that Information Technology has not been fully appreciated, that existing versions of reference databases provide little more than the simple transfer of printed schedule information onto a screen, and are little more than a prelude to their full integration with reservation systems. Only when the sophisticated search capabilities of today´s intelligent databases are integrated with today´s reservation systems will they become tomorrow´s significant work tool. (4)

Where such databases are already linked into multi-access reservation systems, the reference systems can be used automatically to select the most appropriate third party system(s) with which to interconnect. This gives the user the ability to package individual components.

Air schedules are not the only data to be stored in reference databases to be used in this way. Destination, resort, country, visa, health and tourist information is also being included.

Selection Databases. Although databases will aid the selection of components, it is likely that all such reference databases will be product specific. It will be up to the existing packagers of products to develop selection databases that are available on networks, which can in turn can use reference databases and so interconnect with reservation systems.

The first stage in this last scenario has already been started with major UK travel agents creating their own databases on VANs. These are used for a variety of purposes such as tour availability, air fares and client servicing information to directly aid the sales staff in discussion with the client.

For example, UK agents hold details of tours available from a selection of (preferred) tour operators and sales staff look-up this database first when responding to the client´s requirements. The booking is (apparently) made within the agent´s system but in fact unseen by the user, the booking is relayed to the appropriate tour operator for confirmation. In other words a selection database has selected a host reservation system.

With such databases being specific to the packager, travel agents can tailor facilities to their own sales pattern or, equally important, to their client´s requirements. These aspects will probably prove to be the most significant marketing factor in a fiercely competitive market.

Interconnection with Other Projects.

Existing Trends. The history of information technology developments within the travel industry have been a succession of more and more features covering complementary components into comprehensive or linked systems. This trend is likely to continue as more and more services, databases and networks become interconnected.

Travel Packagers. The travel agents or perhaps a new breed of travel packagers will increasingly be able to refer to other systems in order to package products for a client.

As one example of how separate systems might be interconnected to produce a new product, consider the example of a local consortium of hotels, government departments and local suppliers embarking on a project to establish a new conference centre. A number of facilities would be required: central booking facilities for hotels in the area, reference data on hotels, conference facilities (capacities, audio-visual, etc.), trade fare stand contractors, etc.; schedules and booking facilities for local transport; details of and booking facilities for theatres, etc.. All these are likely to be held on a mixture of central and local computer systems. Each component must clearly be coordinated with others. If these individual, but linked, components are now interconnected to international networks, the travel packager could access details, book and otherwise organise conferences at a distance.

Electronic Funds Transfer (EFT). The developments relating to electronic funds transfer are well known and affect everyone. The travel industry with its own very diverse client / supplier transactions are a prime candidate for the full implementation of EFT.

A substantial proportion of all travel business is already transacted via credit and charge cards. Airline and hotel systems already have links for authorisation purposes. The transition to full EFT is (technically) a straightforward matter particularly as

EFT systems are generally based on PSS networks.

Route Guidance and Itinerary Planning Systems. Planning for accommodation and estimating arrival time is necessary, whether several weeks prior to departure or during the trip itself. Up to date knowledge is essential when estimating journey times and route guidance systems may have an important role to play particularly if they can assist the client in avoiding (or the operator in predicting) delays on route. Furthermore the ability to change and optimaly relocate previously arranged stops en-route would allow accommodation to be used to maximum efficiency. Projects such as the European Traveller´s Information System study by the EEC aim to connect transport systems with conventional booking systems. Thus hotel space not now required can be released at the same time as search for alternative accommodation is made. (5)

The interconnection does not have to end there. A link with an Electronic mail system as well could, for example, inform other affected people of the change in plans.

Conclusion. Travel industry information technology has been based on communications technology throughout its history. As developments increase the flexibility and scope of such communications, commercial pressures will force more and more related projects to be combined together to produce an ever more attractive service for the client.

EDITORS' NOTES

(1) See Chapter 13 for a wider discussion of the importance to clients of on line information about fares and services.

(2) The advantages of Personal Computers for local processing functions are alluded to in several chapters in this book. For example, Chapters 5, 7 and 10 describe their role in the rail and bus industries.

(3) Chapter 15 includes a discussion of the role of knowledge based systems in the transport sector and makes particular reference to systems being developed to deal with travel enquiries.

(4) Chapters 8 and 10 allude to the question of attitudes to new computing technology in the air traffic and bus industries respectively.

(5) See Chapter 14 for a review of route guidance systems and Chapter 15 for a discussion of the potential role of expert systems in this area.

Information Technology Applications in Transport, pp. 287–318
P. Bonsall, M. Bell (Editors)

Chapter 13

INFORMATION FOR PUBLIC TRANSPORT USERS

Ling Suen[1] and Tom Geehan[2]
[1]*Transport Canada, Canada,* [2]*TransVision Consultants, Canada*

INTRODUCTION

For a public transport system to be used, the public must know where and when the service is provided. Information is an essential element of a public transport service. Traditionally, network maps and timetable structures have been used in an attempt to provide the necessary information to the general public. These, however, are being superseded by new methods of communicating information to public transport users.

New methods of communicating information to public transport users are of particular interest since the network and timetable structures are so complicated that users encounter difficulties unless they are provided with additional assistance (see Geehan and Deslauriers, 1978 or Everett et al 1977). The need to be informed can only be partially satisfied without the the application of information technology. The decision whether or not to travel by public transport often depends on usable, factual information about the proposed journey at the decisive moment — and this is only possible with the recent developments in information technology and communication systems.

These developments have brought about a major change in the quality of information that can be made available to public transport users. More information can be made available than ever before and it can be transmitted more rapidly, flexibly and diversely. The technical capability is available for fast, effective communication with the general public, providing the public transport user with more usable, factual information.

The capabilities of information technology and communication systems have progressed concurrently over the last few years with an increasing interest in the needs of the general public. This consumer-oriented approach to public transport has developed at the same time as society and public transport operators have become more technology-oriented. The consumer-oriented approach has also been viewed as a means to increase patronage and productivity and thereby improve the efficiency of existing services and facilities.

The consumer-oriented approach recognizes that there are different segments or categories of public transport users comprizing individuals with differing information needs based upon personal, social and economic characteristics and their involvement with the public transport system. The different categories of public transport users include:

- regular passengers, who use public transport for daily work or school trips. Because they are largely familiar with the system, a minimum information requirement can be assumed, and
- occasional passengers, who use public transport for a specific trip purpose, are not very familiar with the transit system and therefore have a correspondingly high information requirement. Occasional passengers include:

  unaquainted passengers, who are strangers in the city,

  exceptional passengers, who use public transit means only in exceptional situations,

  foreign passengers, who have additional communication difficulties and are unacquainted with the local situation, and

  disabled passengers, who have certain functional limitations which may or may not relate to communication difficulties.

The scope of the information required for these public transport users varies from minimal route-specific timetable information to complete trip planning information about how and where to use the service. The method of conveying the information to the user can range from printed materials to interactive video display terminals. The information can take the form of various combinations of symbols, colours, numbers or names. All these components have to be considered when developing new information systems that will meet the needs of the public transport user and each component must be carefully coordinated with the other components.

With this consumer-oriented approach to public transport, the new information technology is of prime interest — relating as it does to human communication processes and to the handling of what is conveyed in those processes; that is, information. Information that can be used to affect individual behaviour. Information that can be used to change attitudes, improve the image of the public transport operation and to increase ridership and revenues.

COMPONENTS

Every information system consists of several components that must be coordinated in their content, form and media. Content, form and media are the irreducible, basic components of all information. The content is primarily what is being expressed, directly or obliquely; it is the character of the information, the message. The message is composed with purpose: to tell, express, explain, direct, inspire, or affect. The form is the method of presentation: the design and arrangement of the basic elements of the message. The media is the location and method of conveyance of the message. It is the interface, or channel of communication between operator and user. Form is affected by content; content is affected by form. The message is cast by the transport operator; and received and modified by the public transport user.

As seen in Figure 1, information systems may be thought of as having three broad levels of content — varying from quite general information to specific schedule and even current real-time operational information. The form of the information can be abstract, symbolic or literal. Various forms are often combined; for example, abstract and symbolic forms, such as colours and pictograms, may draw passengers' attention to more literal information. The media is differentiated by the level of interaction afforded the public transport user — passive, active or interactive. Print is a passive or static type of media; signs and audio signals may be passive or active; and personal interaction is considered interactive.

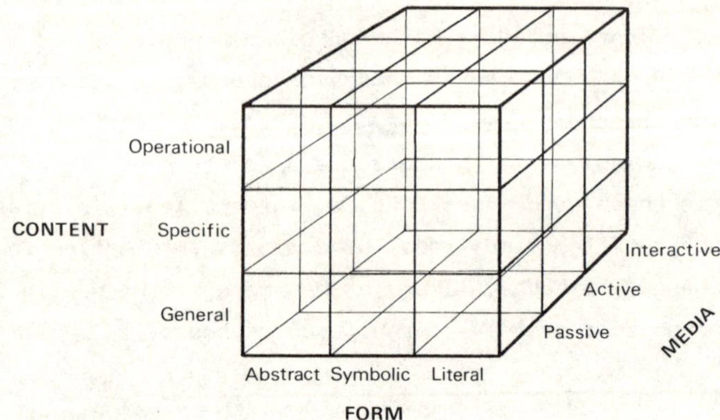

Figure 1: Information Components

Content

In any public transport network, there are several levels of information required by the passengers. The particular level depends on the purpose of the information and tends to dictate its function:

- general information, which exists independently of a specific transit system and refers to public transport in general;
- specific information, which concerns a specific service, including its schedule, fare, payment methods and other conditions; and
- operational information, which concerns the routing and current timing for an actual trip, including connections and emergency instructions.

The information content does not have to be thought of as only operating in distinct levels, but can be expanded in subtle steps on a continuing basis from one level to the other, ranging from basic service instructions to routing, schedule and fare information to detailed technical instructions for specialized equipment, services and facilities, as shown in Figure 2

General information is provided to increase awareness and to improve attitudes towards the transit system. This information is oriented towards the occasional passengers with the ultimate purpose of increasing ridership on the network. Information about the transit system and instructions for obtaining specific or actual information is generally consistent throughout the system and can be treated in a long-range informational or educational manner (Suen, 1982).

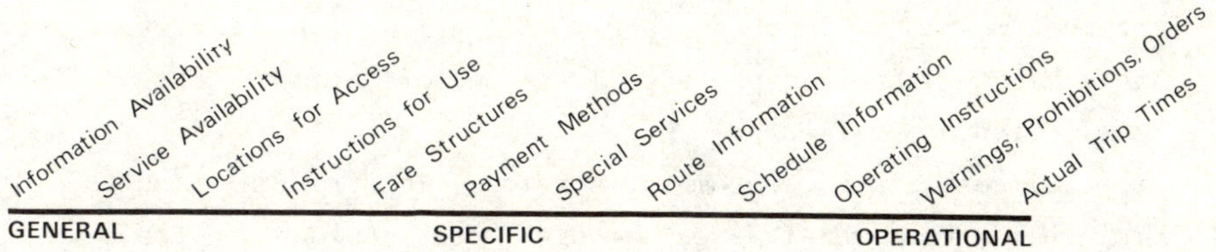

Figure 2: Content of Public Transport Information

Specific information oriented to regular passengers must be provided to ensure that users know where to access the system and how to use it. This includes detailed information for users familiar with the public transport system but who need to know route specific details concerning the system such as is often contained in timetables and maps.

Operational (or real-time) information is provided to public transport users to reduce waiting times, one of the most important deterrents to transit patronage, and to improve the confidence of users in the transit service. The information is oriented toward both the regular and occasional users and can include trip planning or itinerary details. The most important factor with the provision of operational information is that it be provided at the decisive moment for the public transport user.

The information content which must be transmitted to all categories of passengers in an easily understandable form includes the following:

- <u>information availability</u>; where, when and how can the passenger obtain what information?
- <u>service availability</u>, directories to help the user find the desired transit network and routings.
- <u>locations for access</u>, markings to identify the most important

parts of the transit system, such as the stops.

- instructions for use, providing basic information about how to use the system (route, time, price).
- fare structures, determination of the fare according to the fare structure as a function of distance, time and duration of the trip.
- payment methods, purchase and validation of tickets, with information about type and location of ticket sale, fare collection and validation.
- special services, customer service information, containing references to special fares, popular destinations and the like.
- route information, identification of the local transit network with respect to routes, vehicles, access to stops, transfer points.
- schedule information, identification of the departure time and estimated duration of the trip (timetable data). (1)
- operating instructions, explaining the manipulations in the use of technical devices, to be carried out by the passengers themselves.
- warnings, prohibitions, orders, protecting the passenger from dangers due to the operation and the operator from misuse of the equipment.
- actual trip times, identification of the actual (real-time) vehicle departure times from stops.

Individual information as well as the entire information system must be limited to the essential. Too much information is confusing. Lack of information endangers the success of the system. The information provided should guide the passengers through the network as easily as possible. For this reason, information devices must be logically configured for the particular purpose, arranged consistently, and the content and form of the message must be carefully coordinated.

Form

The content of the information (that is, the message) largely dictates the form (the design and arrangement) that the information will take. The provision of warnings, prohibitions or orders will take an entirely different form than instructions for using the service or advertisements to increase ridership.

The basic elements of information combine to form a particular message. The elemental forms contained within the message should be intelligible by themselves. Easily understandable symbols and pictograms are less demanding with respect to intellectual powers than written information and do not presuppose any knowledge of the language by foreigners. Literal forms, on the other hand, are more precise.

The forms of information can be thought of as falling within three distinctive and individual levels:

- abstract forms, which are the understructure or environmental aspects of the message, the compositional forces that reinforce the information experience;
- symbolic forms, which are the representational level of the information, consisting of signs, signals, symbols and pictograms, used to represent specific messages; and
- literal forms, which are the informational material requiring no intravening coded systems to facilitate understanding and no decoding to delay comprehension.

The various forms of information and their levels are illustrated in figure 3:

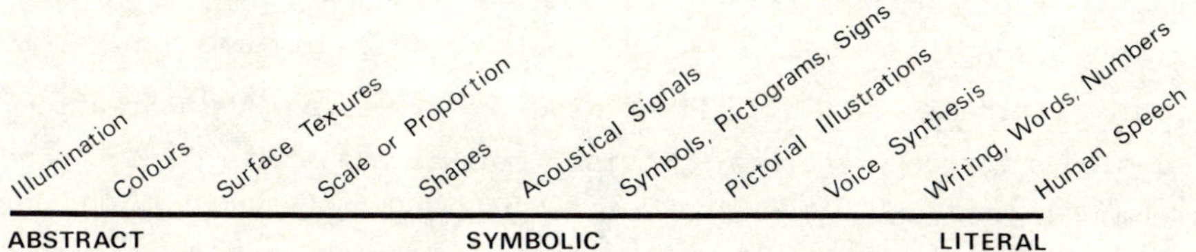

Figure 3: Forms of Public Transport Information

The forms that are commonly used to serve public transport information transmission include the following:

- illumination, which is often used for important information since it improves the legibility and recognizability of the message. Flashing signs, for example, can be used as attention getters.

- colours, the most emotional and expressive element, tend to support the value of information and are often combined with other forms. The colour red, for example, is often used to signify danger.
- surface textures, optical or tactile, the surface characteristics of visual materials, are often used for orientation messages, providing particular aid for disabled passengers (especially the visually-impaired).
- scale or proportion, the relative size and measurement, is often used to indicate the relative importance of the message.
- shapes, dots and lines, such as circles, squares and triangles, are often used as markings to identify components of the system, such as the location of transit stops.
- acoustical signals, including warning signals such as klaxons, sirens, bells, buzzers and attention signals including triad, gong, and buzzer may be used to impart a particular message to the users.
- symbols, pictograms and signs, pictures which are evident, easily intelligible and uniform (for example, pictograms of the International Railroad Association). See figure 4:
- pictorial illustrations, such as photographs, graphs and city maps are appropriate to supplement and explain other information.
- voice synthesis, is often used for repeated information, includes speech storage systems, such as magnetic tapes and computer systems.
- writing and numbers, the most precise and most important method of conveying complex information to public transport users, often used in legends to explain other forms of information in the message.
- human speech, which can provide current operational information either in person (providing the passenger with the opportunity to interact and ask questions) or over a loudspeaker.

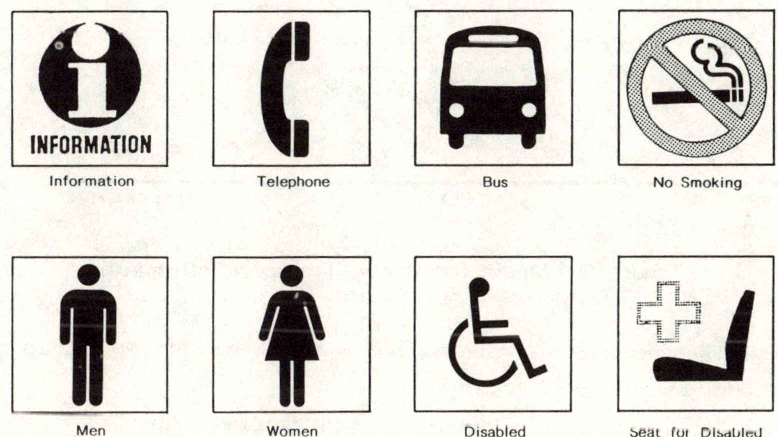

Figure 4: Examples of Symbols and Pictograms

Media

The provision of information or instructions for public transport users can be carried out aurally or visually by humans or various technological devices, (See figure 5). Communication of the information can be with or without the possibility for the passsenger to interact with questions:

- passive media, or static media, in which limited information is conveyed in the same way to all users;
- active media, in which specific and changing information is conveyed to users but which can not be questioned by the user;
- interactive media, in which the passenger can interrogate and from which they obtain the desired information.

Passive media provides static or open communication with no facility for feedback or interrogation. The information is conveyed in the same way to all users. Active media is a more dynamic, but still open, system that also does not provide any feedback to the information transmitter but allows for broader content. Interactive media is media that is dynamic and closed and provides the user with the opportunity for feedback and interrogation. While the possibility for interaction makes the system more expensive, it

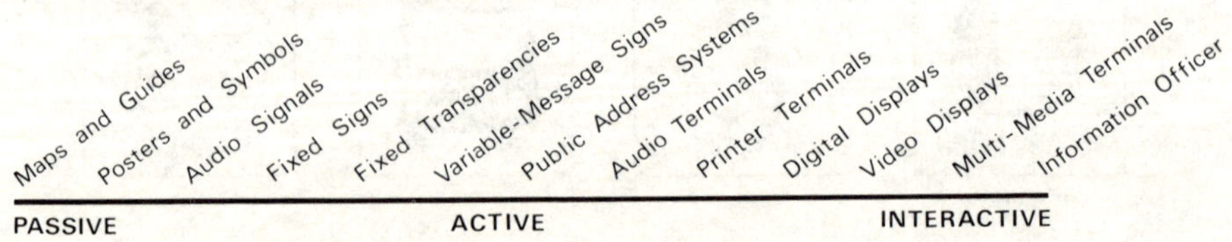

Figure 5: Media for Public Transport Information

allows the passengers to determine whether the information is clear and complete and meets their needs.

The various media for conveying information to public transport users can be delineated in subtle steps on a continuing basis through the various levels from passive to interactive as follows:

- maps and guides, relating the information content to the geographic area. Containing applicable legends, the maps can be drawn to scale or in a schematic presentation as appropriate.
- posters and symbols, providing information or directions in graphic or symbolic form.
- audio signals, such as buzzers and alarms, for arousing attention.
- fixed signs, which are the most frequently used and least expensive type of information transmission.
- fixed transparencies, which are a more expensive form of sign but attract more attention and -- if they are illuminated -- are especially effective during darkness.
- variable-message signs, which are often used for frequently changing information. They are indispensable for remote data transmission.
- public address systems, serving the amplification, dissemination and remote transmission of the human voice or of acoustic signals.
- audio terminals, such as the telephone, providing a means of transmitting human or synthesized voice information remotely.
- printer terminals, which can constantly or intermittently transmit current or required information via print-outs.

- <u>digital displays</u>, which would consist of a more interactive form of variable-message sign, providing simple listings to the user.
- <u>video displays</u>, which can constantly transmit complex information (remotely) via video display units, either actively or interactively. (2)
- <u>multi-media terminals</u>, providing for the transmission of written, visual and audio information interactively.
- <u>information officer</u>, public transport personnel trained and located to provide information to public transport users upon request.

Some characteristics of active and interactive information terminals are outlined in Figure 6. These terminals may stand-alone or they may be connected to a central information centre.

Audio	Booth or wall-mounted Touch-tone or dial telephone Digital and/or voice communications Information provided by voice
Print	Booth or cabinet Simple numeric keypad Information dispensed on printed card
Digital	Kiosk or wall-mounted Colour listings and simple images Keyboard or touch screen
Video	Kiosk or wall-mounted Complex colour images Simple numeric keypad Touch screen
Multimedia	Kiosk style Keyboard Video screen Printed information Synthesized voice messages

Figure 6: Information Terminals

<u>Interaction</u>

In all of its many aspects, providing information for public transport users is complex. Nevertheless, complexity need not be a hindrance in understanding the process. True, it is easier to have one set of common definitions, but simplicity has negative aspects. The simpler the formula, the more limited the potential for creative variation and expression. The three-dimensional functionality of information and communication outlined in the previous section — content, form and media — offers harmonious (3) interaction, syncretistic though it may be. The striking feature of this typology is its simultaneity. Each dimension is linked to the other and interaction between the dimensions is integral.

But the complexity of providing information to public transport users has to be considered in light of the variations involved in its reception. When we perceive a message, we are doing many things at once. We are seeing an enormous field peripherally. We are decoding all manner of symbols. We are imposing on the message our own compositional forces. The reception of information is therefore also a multidimensional process that is influenced and possibly modified by psychological moods and cultural conditioning and by environmental expectations. How we view the world frequently affects what we perceive. The process, is after all, a very individual one for each of us.

APPLICATIONS

In considering applications, the number of choices open to the information problem solver are overwhelming. The options are vast and the forms and media many: the three dimensions of information interact and it is through this interaction that the character of the message takes form. In Figure 7, information content, media and form are related to the function and location of information for public transport users. In reviewing specific applications, we must concern ourselves with form and content, their interactive relationship, as well as the character of the individual elements. In addition, information applications are governed by intended meaning for individual personal and cultural environments. And the last consideration is the medium itself, which through its own character and limitations will often determine the type of application.

Despite the preceding considerations, there is an obvious trend in information systems for public transport users, one that can be used to simplify the complexity of the problem and provide broad

Figure 7: Information for Public Transport Users

CONTENT	MEDIA	FORM	FUNCTION						LOCATION					Requirements		
			General Information	Route Network	Departure Times	Fare Structure	Payment Methods	Service Instructions	On the Vehicle	At the Bus Stop	At Transit Stations	At Activity Centres	Information Centres	Changed Occasionally	Constantly the Same	Legible at ? Metres
Comprehensive information	Documents, maps, timetables, guides	Legend, numbers, colours, illustrations	X	X	X	X	X	X			X	X	X	X		½
Individual information	Phone, info centres, verbal information	Speech	X	X	X	X	X	X			X		X			
General information	Video displays, interactive terminals	Legend, numbers, pictures, symbols	X	X	X	X	X	X				X				½
Special customer service information	Documents, brochures, posters, leaflets	Legend, numbers, symbols, colours	X	X	X	X	X	X	X		X		X	X		½
Instructions for information availability	Sign	Pictogram ('information') with arrow	X								X				X	20
Reports and announcements	Press, magazines, radio, television	Legend, photographs, speech, numbers	X										X			½
Location of routes and bus stops	Maps	Legend, numbers, colours, symbols		X							X		X	X		½
Marking of bus stop position	Sign/transparency	Symbol, shape, colour		X						X	X				X	00
Name or number of the bus stop	Sign/transparency	Legend, number		X						X	X				X	20
Routes served by the bus stop	Sign, display	Legend, numbers, graphs		X						X	X			X		5
Route number with direction of operation	Sign/transparency, variable message sign	Legend, number		X					X					X		50
Indication of transfer/ route locations	Voice and/or public address system	Speech		X					X							
Description of route or itinerary for trip	Automatic destination selection devices (ATIS)	Legend, optical signs		X	X	X					X			X		½
	Telephone information service (AIDS, CCIS)	Speech		X	X	X							X	X		
Route timetable	Documents	Legend, numbers			X						X		X	X		½
Time	Variable message sign	Numbers (digital or analog clock)			X						X	X	X		X	50
Arrival or departure time at the stop	Variable message sign or sound (PIBS, ELSIE)	Legend, numbers, or speech			X					X	X			X		5
	Phone/voice synthesis (ERICA, TELERIDE)	Speech			X								X	X		
Warning before departure	Attention signal transmitter	Optical or acoustical signals			X				X					X		10
Indication of next bus stop	Attention signal transmitter	Optical or acoustical signals			X				X					X		10
Fare system	Documents	Legend, numbers, colours, graphs				X			X		X		X	X		½
Directions for ticket purchase/validation	Sign	Legend, pictogram					X		X		X			X		½
Fare payment procedure	Sign	Legend, pictogram					X		X		X			X		1
Door operation and exit instructions	Sign	Legend (with arrows), pictograms, colours						X	X						X	½
Breakdown information	Loudspeaker, variable message indicator	Signs, sounds, speech						X	X					X		10
Instructions in case of emergency situations	Voice or radio telephone system	Speech						X	X					X		

Source: Adopted from SNV (1982) p132.

categories for analysis, and this trend is towards improved availability of information through the application of new technology. The transference of information processing from trained personnel to technological devices not only reduces personnel costs but allows for greater user accessiblity through a broader placement of relatively inexpensive media. At the same time, improved communications systems allow for economies inherent in the centralization of information storage and retrieval devices.

The complexity of information systems does not therefore defy definition. The trend resulting from the application of new technology illustrates a number of possible categories of information system defined according to the source or location of the information as perceived by the public transport user:

- on the vehicle, by means of visual signal systems to identify the vehicle and its routing and by communication with the bus driver.
- at bus stop, where various visual or acoustic signal systems may be placed for providing location and/or time information.
- at transit stations, where specific and operational information can be provided through a greater variety of signs and displays.
- at activity centres, where selective information may be provided through interactive terminals in a controlled environment; and
- from information centres, where the telephone is a means of providing a large amount of information to public transport users.

(4)

The following applications of information technology to the provision of information for public transport users are presented according to these categories. These modern information techniques, which are summarized in Figure 8, enhance the quality and availability of information to public transport users. The techniques are outlined in the figure according to the content of the information provided, the media used for accessing and disseminating the information, and the cost and status of the applications development effort. The applications are outlined in greater detail in the following paragraphs. Although specific individual applications of particular interest to public transport

Figure 8: Information Technology Applications

SYSTEM	CONTENT	MEDIA			STATUS	COST
		Type	Access	Output		
BRISTOL ELSIE	Schedules	Active	Push Button	Synth Voice	Demonstration in Weston-Super-Mare	£7,000 for prototype
NICE VIDEOBUS	Real-Time	Active	Display	Digital Display	Operational in Nice since 1981	FF 9,000,000
LONDON PIBS	Real-Time	Active	Display	Digital Display	Under test in London	n/a
WUNSTORF RETAX	Trip Details	Active	Call Terminal	Printed Voucher	Operational in Wunstorf	n/a
FRED'SHAFEN RUFBUS	Trip Details	Interactive	Call Terminal	Printed Voucher	Operational in Frederichsafen	n/a
COLUMBUS TOUCHMEVISION	Routes Schedules Fares General	Interactive	Touch Screen	Video Display	Operational in Columbus, Ohio	US$15,000 per unit
TELERIDE VIDEO	Routes Schedules Fares General	Interactive	Keypad	Video Display	Under Development in Duluth, Minnesota	US$10-50,000
PORTLAND KIOSKS	Routes Schedules Fares General	Interactive	Keypad	Video Display	Operational in Portland, Oregon	n/a
SAN FRANCISCO AGTIS	Itinerary Schedule Fares Trip Time	Interactive	Keypad	Video Display	Operating at San Francisco Airport	n/a
MISSISSAUGA, CANADA ERICA	Real-Time	Active	Touch-Tone Dial Phone	Synth Voice	Now provides only schedule information	C$1,260,000
HALIFAX, CANADA GO TIME	Real-Time	Active	Touch-Tone Dial Phone Display	Synth Voice Video Display	Operational in Halifax, Canada	C$1,350,000
TELERIDE ATIS	Schedules	Active	Touch-Tone Dial Phone Display	Synth Voice Video Display	Operational in in 10 Cities in US and Canada	C$400-700,000
HAMBURG ATIS	Itinerary Schedule Fares Trip Time	Interactive	Touch-Tone Dial Phone Disp Terminal	Synth Voice Screen List Printed Card	Operational in Hamburg	US$1,000,000
WASHINGTON AIDS	Itinerary Schedule Fares Trip Time	Interactive	Voice Phone Term Operator	Video Display Voice Phone	Operational in Washington, D.C.	US$1,000,000
LOS ANGELES CCIS	Itinerary Schedule Fares Trip Time	Interactive	Voice Phone Term Operator	Video Display Voice Phone	Operational in Los Angeles	US$1,500,000
DENVER TRIPS	Itinerary Schedule Fares Trip Time	Interactive	Voice Phone Term Operator	Video Display Voice Phone	Under Development in Denver, Colorado	US$540,000
MANCHESTER TRAVELGUIDE	Itinerary Schedule Trip Time	Interactive	Key Pad	Printed Card	Under test in Manchester, England	n/a

undertakings have been selected for illustrative purposes, they should only be considered indicative of the alternatives possible.

On the Vehicle

On the outside of the vehicle, information for public transport users includes the route number and destination of the vehicle. This information is displayed in large, bold numbers and letters on the front of the vehicle, on the passenger side, and, desirably, on the rear of the vehicle. The lettering needs to be large and easily visible by waiting passengers under possibly adverse weather and lighting conditions. The information is further affected by poor angles of view, obstruction by obstacles, adverse illumination, changing distances and other factors. As a consequence, public transport undertakings have been investigating new technological alternatives, such as variable-message digital destination signs, to provide more legible information on the outside of vehicles for public transport users (Figure 9).

The simplest method of applying outside information is the inserted sign, which is normally used only for special destinations or when the destination has to be rarely changed. The most common information system is the rolling sign indicator. The rolls can include up to 120 destinations and are translucent. Back-lighted white symbols on a black background have proved to be the most recognizable. Quick change rolling band cassettes increase the number of destinations. The rolling bands can be activated manually or by means of a motor (see SNV, 1982).

Flap panel indicating devices, which have been introduced successfully as stationary signs, have not proven adequate as destination indicators on vehicles. The flap panels require much more space, are not transparent, are difficult to illuminate, and their legiblility is generally poor, especially in sunlight. The required height for the lettering allows for only about 40 destinations to be stored in the display and the plastic panels tend to flap down by themselves due to vibrations on the vehicle.

Digital display systems are the most versatile for destination

Figure 9: Destination Signs

DESTINATION SIGN	Type of System	Recognizability During the Day	Recognizability at Night Time	Side View	Contrast	Legibility	Number of Destinations	Space Requirement	Weight	Energy Requirement	Reliability	Service Life	Operating Costs
Inserted Sign	Passive	Good	Good	Good	Good	Good	Limited	Low	Low	Low	Good	Long	Low
Rolling Sign	Partially Active	Good	Good	Good	Good	Good	120	Justifiable	Justifiable	Low	Fair	Long	Average
Flap Panels	Partially Active	Good	Good	Fair	Good	Good	39	Greater	Could be less	Low	Fair	Long	Low
Projection	Partially Active	Not Good	Not Good	Poor	Poor	Poor	Optional	Too Great	Justifiable	Low	Poor	Long	Average
Field of Lamps	Active	Not Good	Good	Poor	Poor	Satisfactory	Optional	Considerable	Too High	Greater	Poor	Average	Average
Liquid Crystal (LC)	Active or Passive	Not Good	Good	Fair	Poor	Unsatisfactory	Optional	Justifiable	Too High	Low	?	Still too Short	?
Ferro-electric Bistable Displays (PLZT)	Active or Passive	Not Good	?	Poor	Poor	Unsatisfactory	Optional	Justifiable	Too High	Low	?	?	?
Light-emitting Diode (LED)	Active	Not Good	Good	Fair	Poor	Unsatisfactory	Optional	Justifiable	Justifiable	Low	Fair	Still too Short	?
Cathode Ray Tube (CRT)	Active	Poor	Not Good	Poor	Poor	Unsatisfactory	Optional	Too Great	Too High	Considerable	Low	Too Low	High
Plasma Screen	Passive	Not Good	Good	Good	Fair	Satisfactory	Optional	Justifiable	Could be less	Justifiable	?	Average	High
Electro-chemical Display	Active or Passive	?	?	?	?	?	Optional	?	?	?	?	?	?

Source: SNV (1982) p102.

indicators. A large variety of alternative technologies, from liquid crystal to plasma displays (see Graff, 1984) can be considered. These are extremely flexible and, with the aid of changeable individual elements and software programs, any number of texts can be written. Widespread application of these systems, however, is contingent upon improvements in the legibility of the displays and, at the same time, ensuring that the energy consumption is within reasonable limits.

Information inside the vehicle should comprise information about the next bus stop and transfer possibilities, information about the assurance of connections, and announcements about emergency situations.

Currently, information in the bus is often distributed over the entire vehicle. Confirmation of the passenger's stop request, for example, which is initiated through a push-button or pull-rope, may be given through a transparent sign reading 'Next Stop'. Other signs or brochures may relate to the fare structure, route structure or operating instructions. Announcements of the bus stop or transfer possibilities by the driver are often omitted or cannot be understood. Information about the assurance of connections or emergency situations is rarely provided.

Although the desired information may be provided by the bus driver with a public address system and a tape recorder, it may also be provided by a central control centre in a transit system that has a computer-controlled communications system. In emergency situations, it is possible to transmit information from the control centre, through loudspeakers, to the passengers in the vehicle. For example, Toronto's Communication and Information System (CIS) contains a public address component that can be accessed by either the driver of the vehicle or the control centre personnel (described by Peat Marwick and Partners, 1976).

At Transit Stops

At less-frequented transit stops, only brief bus stop information is provided for public transport users. This may include: marking of the stop location, name of the stop, route number and indication of direction. This information may be expanded, in some instances, to include a route timetable with information about bus stops on the route and scheduled departure times, as well as indications of the fare system and fares, ticket purchase and validation.

Bus stop information can be expanded considerably by transmitting operational information such as, for example, the arrival of the next bus or the cause and duration of breakdowns, to the waiting passenger. A prerequisite for the provision of current operational information is operational monitoring and control by means of a computer-controlled operations control system.

An example of providing information at transit stops is the talking stop developed by the Department of Transport in the United Kingdom. An acoustic signal is used to help the disabled to locate the stop. The stop can then be manually activated to speak the route number of each service using the stop and to give the scheduled arrival time of the next bus on each route. The talking bus stop has been given the name ELSIE, for 'Electronic Speech Information Equipment'. A field trial in the town of Weston-Super-Mare in Avon is currently in preparation (see Hobbs and Smallwood 1984). Initially some nine bus stops in and around the town will be equipped.

London Transport is also currently evaluating a prototype Passenger Information at Bus Stops system, known as PIBS. The system provides passengers waiting at bus stops with information about when and which of the buses serving the stop will arrive next. A visual presentation, using a bright illuminated dot matrix display, rather than an audio message is used. The display provides a countdown to the time of arrival of the next bus. The system is being tested preparatory to its wider introduction along major bus routes in the British capital (see UTA, 1984).

The London Transport system is similar to the one that has been used in Nice for five years. The Nice VIDEOBUS system uses a display at the bus stop to indicate the routes serving the stop and indicator lights to show the position of buses on the routes. In this way, passengers can follow the progress of the bus they are

Illustrative Applications

Bus passenger display, MBB

Mini call terminal, Retax system

Call terminal, Retax system

Trip request via call terminal

Call terminal, Rufbus system

Bus stop display, Nice Videobus

Illustrative Applications (cont'd)

Automatic information, MBB

Automatic information, Hamburg

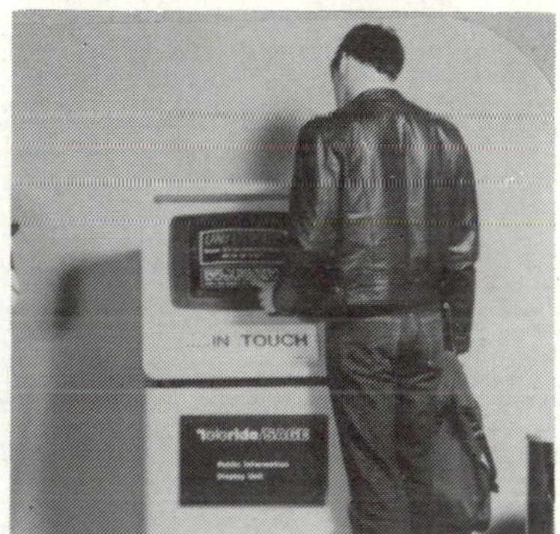

Interactive video, Teleride

Interactive video display, TFP

Video schedule display, Ottawa

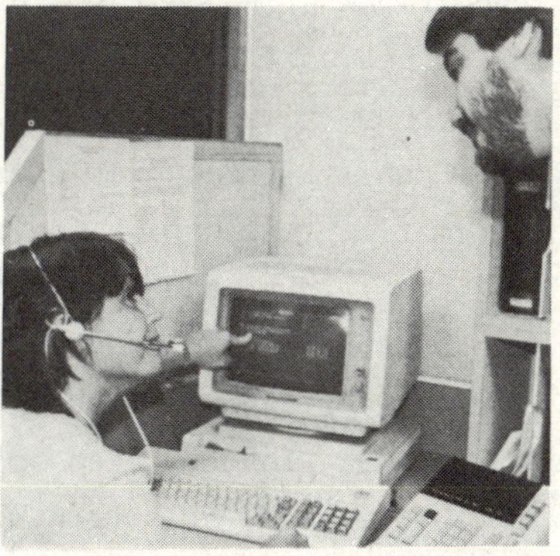

Information service, Los Angeles

waiting for. For reasons of size and legibility, the capacity of the boards is limited to three routes with about 10 indicator lamps each. In view of the success of the operation, the City of Nice decided in 1983 to extend the VIDEOBUS system along several major routes (see Artaud-Macari 1983).

Within the scope of research on the use of on-demand bus systems, the possibilities for a dialog between the passenger from the bus stop and the control centre are also being explored in several cities. In these systems, passengers can report their trip requests to the control centre through either the telephone or special call terminals. In the case of trip requests through call terminals, the passengers key the code number of the stop and the number of persons and obtain information about the vehicle number and expected time of arrival, which is indicated on displays at the call terminal and/or by means of printouts. Examples are the systems RUFBUS (at Fredrichshafen) and RETAX (at Wunstorf) - see Meyer (1982). (5)

At Transit Stations

At frequently used bus stops, especially at transfer points, detailed information should be provided about public transport routes, timetables, the fare system and fare. In addition, displays can be erected in connection with shelters, suitable for accomodating transit route maps, maps of neighbourhoods or supplementary information concerning the fare system and the conditions of transit. The marking of the stopping point, as well as directional signposts or a clock could be included.

At these heavily used bus stops, which are served by many routes, the public transport user can be further informed by means of automatic destination selection machines that provide passengers with individual route descriptions. Existing systems use the city map with an additional listing of the bus stops in alphabetical order (for example, the Metro in Paris or the city transit in Cologne). Input takes place either indirectly, through the activation of a button on the city map or in a bus stop list or indirectly, through a code number assigned to the bus stop, which must be keyboarded by the passenger. Following input of the destination and possibly the desired departure time, the automatic destination selection device determines the most advantageous route. The subsequent readout, possibly as a printout, provides information concerning the appropriate routes, transfer points and departure times.

A further example was developed on behalf of the Federal Ministry for Research and Technology in Germany. It provides information necessary for reaching the destination, concerning routes, transfer station with arrival and departure times, as well as fares. The information device was developed by Messerschmitt-Bolkow-Blohm GmbH in cooperation with the Hamburger Hochban AG and was introduced to the public in the summer of 1978. A ten-digit keyboard is used to input a destination code which is selected from a list by the user with the aid of a microfilm reader. After the best itinerary is calculated, the device prints out the recommendations for the journey with all the necessary information. An initial survey revealed that more than 90% of the users could understand the recommendations issued by the machine. In view of the initial success, the number and capabilities of the terminals are being expanded and an associated telephone information service using voice synthesis apparatus has also been expanded (see Grabe et al 1980).

Another similar concept is the Automated Ground Transportation Information System (AGTIS) developed by the United States Transportation Systems Center at Cambridge, Massachusetts. Its prototype featured a touch-sensitive video terminal to facilitate input to a computer-based information retrieval system. The customer uses the touch screen terminal to indicate his destination and then receives visual information to that destination. A working example of this concept can be found at the San Franscisco International Airport. Realizing that complete, accurate, and readily available information is one of the keys to inducing greater use of existing public transport services, airport staff helped to implement a user-interactive, computerized Ground Transport Information System in early 1984. The system includes a small keyboard and colour video display. After selecting from a list of destinations, the user is presented with the costs and travel times of reaching the chosen destination by various modes of transport. Once a mode is selected, more detailed information on schedules and transfer requirements is presented. A printout of this information is then produced for the traveller (see Fein, 1984).

At Activity Centres

Information can also be provided at activity centres, shopping centres and downtown malls. Generally, the information provided in these locations is of a more general content and may be combined with other hospitality and tourism information. Although public transport service availability and directions for use are usually the main messages, more specific information, such as schedules from the centre to various destinations, may also be provided.

Four self service computer terminals for the provision of bus route and schedule information, for example, have been installed at each of two Washington (DC) suburban shopping centres. The terminals present a list of choices to the user including restaurants, stores, theatres, and services. Following selection, specific information appears on the screen. Bus service information includes schedules from the centres to various destinations, explains the routes, provides fare information, and lists useful telephone numbers (see Consultancy Briefing 1,11).

The transit authority in Ottawa, Canada, has installed video displays of schedule information, using the same data base and the same software, slightly modified, as its automated telephone information system (described below) . The first two video installations were made in late 1982 at a local shopping centre and at the intercity bus terminal. The displays show the route number and name, and the number of minutes until the next two arrivals at the stop. Within the next five years, the transit authority expects to have the video displays at a number of centres (see Whelan, 1983).

Instead of providing a telephone information service, the public transport operator in Portland, Oregon, installed eight trip-planning kiosks at the downtown transit mall in 1978. These kiosks were equipped with transit system maps and a television screen where customers could find answers to scheduling questions for any route just by pushing buttons. This system displays graphic as well as alpha-numeric information.

The Central Ohio Transit Authority has taken this concept further. Their system uses full motion picture video with text and voice. The system, developed by Technology for People (TFP), is activated through a touch-sensitive screen rather than the traditional key pad and a built-in printer enables the user to walk away with helpful information in hand. Another system, but with keypads, is being developed by Teleride for the City of Duluth, Minnesota. Keypads were selected in Duluth due to the concern for vandalism costs if the screens were easily accessible. Similar systems were used at Expo 85 in Japan and IBM is installing another system at Expo 86 in Vancouver, Canada.

From Information Centres

Passenger information from a public transport authority's central information service has typically been provided by telephonists relying on their knowledge of the system as well as printed information on specific routes. The main problems with this approach are the length of the time before an operator can be fully trained and useful, and the difficulty of organizing printed materials for easy access. Two levels of technology have been implemented to try to alleviate these problems. (6)

The first level involves the use of computers to store the route information for easy retrieval by the telephonists. With such systems, access time is reduced and significant gains in telephone operator productivity can be achieved. Two notable systems have been developed in the United Stated, namely the Computerized Customer Information System (CCIS) developed for the Los Angeles transit authority (see Hebert 1985 a, b) the Automated Information directory system (AIDS) by the Washington transit authority (see

Page, 1981). These are basically telephone information systems providing callers with their optimal point-to point itinerary once the desired travel date, times and origin and destination information is provided. Both systems use manual call-in and answer-back procedures; that is, the caller has to dial in and enquire verbally and the call-taker responds by retrieving information via a computer terminal and then reading the information over the telephone. Both systems have been in operation since about 1980. A similar (but not operational) system, called Travel Guide, was developed by Wootton Jeffreys in 1981 using data from the Bury area of Greater Manchester, UK (see Williamson and Miller, 1981).

The second level of information technology makes use of digital communications and voice synthesis. The telephone recording of trip requests by personnel in the control centre is especially user-oriented, but relatively expensive. Cost savings can be realized when it becomes possible to use the telephone as a data input device. The information can then be transmitted to the user by means of synthetic speech.

In Hamburg, route planning information is provided in this way. The user must first find the reference numbers of the two addresses he wants to travel between by referring to a printed index of addresses. He then rings the information service and gives the reference numbers of the departure and destination addresses, the day of the week and the preferred time of departure by means of the telephone dial. The instruction to dial the trip requirements, the confirmation of the details and the appropriate recommendation are all transmitted to the user by the computer in synthetic language (see Grabe, 1980).

Another application of computerized voice response is simply to provide telephone callers with the times of the next two buses serving the route and stop number specified by the caller. These Automatic Passenger Information (API) systems use voice synthesis technology to answer telephone enquiries for scheduled bus arrival times , service status and other information of interest to the public. They provide prospective transit users with scheduled bus arrival times at their specific bus stops. All bus schedules are stored in a computer along with a directory of bus stops. Each stop is assigned a telephone number. When a call is received, the computer automatically associates the call with the appropriate bus stop, and then responds with a voice message giving the arrival times of the next two or three buses at that stop. At important passenger boarding locations, video monitors can be installed to provide visual information equivalent to that provided by the telephone system.

The concept of the automatic passenger information system was conceived and developed by Transport Canada and a demonstration service was inaugurated in Mississauga, Ontario in 1975. This demonstration project, called Easy rider Information with Computerized assistance (ERICA), illustrated the many benefits of improved information for public transport users (see IBI and Kates, 1979). After the initial success of ERICA, Teleride Developments Limited of Toronto, was formed to market and provide automatic passenger information systems to other transit properties. At the present time, automatic passenger information systems are operating in eleven cities: Ottawa, Toronto, Winnipeg, Mississauga, Kitchener, Brantford and Guelph in Canada; and Columbus, Fort Wayne, Pittsburgh and Salt Lake City in the United Stated. Other cities that are developing their own systems for automatic passenger information include Hull, Quebec and Halifax, Nova Scotia (see Frederick, 1985). Both these systems provide real-time information, similar to ERICA.

One of the more successful installations is in Ottawa. In 1980, the transit authority in Ottawa introduced the automatic passenger information system in one part of the city. Ridership was carefully monitored in both a 'test' community and a 'control' community. The minimum ridership increase due to the system was found to be 2.8% overall and 8.2% in the off-peak periods. Extrapolation of these figures to the whole service area showed that the system would more than pay for itself in the first year. The estimated installation cost for the whole service area was C$0.66 million and the estimated revenue due to additional riders was C$1.11 million in the first yeasr. Based on these results, the transit authority concluded that automated passenger information must be a key component in future ridership growth in the off-peak periods since it provides the necessary customer information in sufficient detail and quantity at an acceptable cost.

IMPLICATIONS

This necessarily short review of the application of information and communication technologies to the provision of information for public transport users may give the impression of extreme diversity and singular accomplishments. This would be misleading. Any individual application in public transport cannot be divorced from another.

The technologies used for conveying information to users are really multiple technologies, since they complement each other, and an investment in one technology often leads to additional commitments and benefits in others (see Transvision 1984). Each technology can show widespread 'ripple' effects throughout the public transport service.

The benefitting applications are also interdependent. An interactive information system providing real-time operational information to the user is not possible without a corresponding central monitoring and control system. The technologies used in both systems complement each other. The information system for the user may even be, and often is, a byproduct of the overall operational system.

(7)

The requirement in public transport operations for more information, particularly information of a highly complex nature, together with the high costs of accessing and processing it in manual form, has created a demand that has effectively complemented the technical push evolving from accelerating developments in technology. The demand is evident and the technical capabilities are considerable.

The real issues affecting the application of information technology to public transport information systems are not technical, but institutional and, to some extent, economic.

Information systems that have benefited from the incorporation of new technology have done so not because of the exciting technical aspects involved, but because of the benefits they appear to offer. The major reason for adopting these technologies are related to economics: the enhanced system capability; the potential for manpower savings; and the quality of the information provided. The systems provide vital and timely information to the users and valuable managerial information to the transport undertakings. They help users reduce uncertainty and result in substantial increases in ridership. although the systems are justified on economic grounds, we must recognize that it is difficult to quantify the benefits, especially since they are only just beginning to be understood. Some systems have experienced a quantifiable benefit, but it is too early to say if this will be sustained.

With the increasing societal orientation to information technology, there is little doubt that continuing varied applications will be found in the public transport industry. The speed of diffusion of these applications will depend primarily, not upon technical or economic factors, but upon human factors. It is not a question of what can be achieved from a technical standpoint but of what can be achieved from an institutional standpoint.

The determining factor in the emerging pattern of technological applications for public transport users is not what can be done, but what is chosen to be done. Technically, a vast range of capabilities are already available, and only the minority of technically demanding applications even approach, let alone are constrained by, the remaining technical limits. Economically, the cost-effectiveness of many of the systems have been proven. The constraining factor is the vision of management in understanding and selecting the appropriate applications and the availability of skilled staff to support their operation.

As seen, many motivations justify the interest in the application of technology to the provision of information for public transport users. These include: (1) the integration of these systems with other central monitoring and control systems for public transport; (2) the ubiquity and quality of the technology and its capacity to facilitate interaction and transaction; (3) the cost-effectiveness of the technology in this particular application; and (4) the acceptability of the systems within a society that is becoming increasingly technology-oriented. Major barriers to adoption of such systems include the ecomonic, psychological, and institutional investments in existing systems.

Nonetheless, in spite of the barriers, new systems will continue to be developed as operators attempt to improve the availability and timeliness of information for public transport users. As the systems become more widespread, the needs of public transport users to be informed will be increasingly met in a more cost-effective manner. The new information and communication technologies are making information more readily available and more useable by the public than ever before. The public transport user no longer has to be content with limited information.

REFERENCES

Artaud-Macari, J, (1983) "Nice: Development of the VIDEOBUS and CIBUS Systems". UITP Review, 32, 1.
Consultancy Briefing "Information Terminals". Consultancy Briefing 1, 11.
Everett, P B et al, (1977) "Transit Route Pamphlets: Do They Work?" Transit Journal, 3, 3.
Fein, S R, (1984) "San Francisco International Airport". ITE Journal, June 1984.
Frederick, S, (1985) "GoTime' in Halifax". Metro Magazine, 81, 2,
Geehan, T and Deslauriers, B, (1978) "Transit Route Guides: An Investigation and Analysis". Bus Ride, Sept 1978.
Grabe, W et al. (1980) "Development and Testing of an Automatic Timetable Information System (ATIS) for Public Transport in the Area Served by the Hamburger Verkehrsverbund". UITP Revue, 29, 1.
Graff, G, (1984) "Liquid Crystals: Big, Bright, Even Colourful Displays". High Technology, May 1984.
Hebert, R. (1985a) "LA's TRANSMIS - The Tool for Modern Management". Mass Transit, 12, 7.
Hebert, R. (1985b) "Los Angeles Pioneers Bus Passenger Info". Mass Transit, 12, 9.
Hobbs, A and Smallwood, T, (1984) "ELSIE, A Talking Bus Stop". Presentation at the 15th Annual Public Transport Symposium, University of Newcastle upon Tyne, England. April 1984.
IBI Group and Josef Kates Associates Inc. (1979) "the Mississauga Transit demonstration of the automatic bus passenger information concept". Prepared for the Urban Transportation Research Branch, Transport Canada, Montreal.

Meyer, H H, (1983) "Operation of Computer-Controlled On-Demand Buses in the Federal Republic of Germany". UITP Revue, 31, 3.
Page, E, (1981) "Automated Transit Information System". Transit Technology Briefs 1, 3.
Peat, Marwick and Partners, IBI Group. (1976) "Communications and Information Systems Development Program". Four reports prepared for the Toronto Transit Commission and the Ministry of Transportation and Communications, Toronto, Canada.
Studiengesellschaft Nahverkehr mbH. (1982) "Bus Transit System: Vehicle Roadway Operation". Verband offentlicher Verkehrsbetriebe (VOV) and Verband der Automobilindustrie (VDA), Germany.
Suen, Ling. (1982) "Automated Bus passenger Information: A Transit Marketing Tool". Presentation at the 12th Annual Public Transport Symposium, University of Newcastle upon Tyne, England. April 1982.
TransVision Consultants Ltd. (1984) "Microelectronics and Urban Transit". Prepared for the Transportation Development Centre, Transport Canada, in cooperation with the Canadian Urban Transit Association, Toronto, Ontario.
UTA (1985) "New Passenger Information System". Urban Transportation Abroad, 8, 3.
Whelan, M S (1983) "Video Schedule Displays Improve Public Information for Passengers". Bus Ride, May 1983.
Williamson, R H and Miller, A J, (1981) "A Personalised Public Transport Directory". Presentation at the 13th Annual Public Transport Symposium, University of Newcastle upon Tyne. July 1981.

EDITORS' NOTES

(1) An indication of the way in which the production of timetable information may be automated and linked to the timetabling operation itself is given in Chapters 7 and 10 for the rail and bus industries respectively.

(2) For further information on the distinction between different types of video system, see Chapter 1.

(3) Syncretism: an attempt to unify or reconcile incompatible schools of thought.

(4) An alternative classification, in the context of information for drivers, is suggested in Chapter 14.

(5) An extension of the concept behind such systems (demand responsive services with a strong emphasis on the provision of information to potential users) has led to suggestions for on line passenger transport brokerage systems. By means of such systems a client requesting a particular journey would be offered a selection of the available modes and services and, in the light of the demands logged, the supply of services could be progressively better matched to the demand for them.

(6) The role of expert systems to overcome such problems is described in Chapter 15.

(7) An example of the joint development of operator and user information systems outside the sphere of public transport can be found in Chapter 6, where the links between road traffic control and the provision of information to drivers are discussed.

Information Technology Applications in Transport, pp. 319–351
P. Bonsall, M. Bell (Editors)

Chapter 14

ROUTE GUIDANCE AND IN-VEHICLE INFORMATION SYSTEMS

David Jeffery
Transport and Road Research Laboratory, UK

INTRODUCTION

Public investment in the road network has decreased in recent years, while traffic has continued to grow. There is therefore an increasing need to use existing roads more efficiently. One possibility is that the continuing advance in information technology may provide new systems for informing drivers about road conditions and route availability and so encourage better utilisation.

Some measure of the factors thought to be most important by drivers in their choice of route was obtained by Andrew (1977). In this survey the top ranking factors out of 18 put to the drivers were (in rank order): total time spent on journey; ease of following route; ease of planning route; condition of road surface; and number of times you are stopped. The likelihood of being involved in an accident was rated as seventh in importance and when asked which hazard warnings would be most useful to them, 'fog' and 'accident ahead' were thought more important than 'congestion' or 'bad weather'.

Much of this information is already available at urban traffic and police control centres, in local authority offices and from the motoring organisations. Information on congestion, weather conditions, traffic incidents and changes in the road network brought about by new construction and maintenance is sought by police and the motoring organisations and efforts are made to use this to the best effect. The main problem lies in the field of information transfer; ie, of collection, collation, and dissemination - and it is here that IT offers great potential.

BENEFITS FROM IMPROVED GUIDANCE AND INFORMATION SYSTEMS

The benefits if drivers' needs could be satisfied are often intangible and many cannot be quantified. Thus it is difficult to estimate the value of increased reassurance to drivers that traffic delays will be minimised and that they would be informed in good (1) time of incidents likely to affect their journey. Nevertheless when looking for possible solutions to problems it is usually wise to consider what the potential benefits are worth and whether they are likely to outweigh the costs.

Observations show that drivers on average tend to behave in such a way as to try to minimise the generalised costs of their journeys, ie, to trade off time against distance, see for example Jeffery (1981). Analysis of times and distances involved in a sample of journeys made in the UK showed average inefficiency to be about 6 per cent. Of this wastage, about one-third was contributed by drivers on journeys for scenic or other reasons which would be unlikely to benefit from an improved guidance system, but the remainder was contributed by drivers who stated they were actually seeking minimum time or distance routes. However, many of the journeys which contributed to the inefficiency were on routes familiar to the driver such as commuter or short shopping trips and it was felt that, since these journeys were made repeatedly, the routes were likely to have been chosen by the driver for some personal reason. These journeys were therefore discounted but even then it was estimated that about 2 per cent of all journey costs remained and these in principle could be recovered by an efficient route guidance system.

For the typical private car driver the wastage is incurred on about 60 journeys each year which are longer than 5km in length and take place over unfamiliar roads. On these journeys the average inefficiency is about 20 per cent and gives rise to additional cost averaging 70p per trip, ie, a total of £40 per year. For the average commercial vehicle the avoidable wastage arises on about 45 journeys per year where excess costs averaging 25 per cent and £3 per trip make a total of about £130 per year. In UK national terms it would seem that the benefits recoverable by an efficient route guidance system would be about £780M per annum of which about 15 per cent is contributed by direct fuel costs, about 40 per cent from other vehicle running costs, and the remainder from a valuation of vehicle occupant's time. If it is assumed that accidents and road maintenance are in proportion to vehicle mileage then a further £44M a year may be available from reduced accidents and £43M from road maintenance thus bringing the total benefit figure to above £860M a year.

The UK Transport and Road Research Laboratory (TRRL) have estimated the benefits of a system that could assist drivers by conveying information about congestion caused by significant traffic incidents. (In 1976, such incidents, causing at least 12 minutes of delay, were estimated to occur 17,000 times a year on UK roads - nearly 7000 on urban roads, about 10,000 on rural roads and about 300 on motorways.) An information system enabling a significant proportion of drivers to divert and avoid such traffic incidents would yield benefits of between £10M and £20M per year at present-day values.

But such a system would still leave drivers on familiar trips, eg commuters, following routes largely chosen by themselves. Analysis has shown that such trips incurred some 2 per cent of wastage in their own right. Additional significant benefits worth up to a further £800M or so per annum may therefore be available to a fully traffic responsive route guidance system if it could influence drivers on familiar as well as unfamiliar journeys. COST 30 bis (1985) suggests that the total attainable benefits could be worth around £4,000M p.a. for Europe as a whole.

METHODS FOR DISSEMINATING INFORMATION TO DRIVERS

There are several possible methods for disseminating route guidance and traffic information to drivers, including for example, maps and guides, radio and TV broadcasts, road signs, and special electronic systems. They are classified into one of five classes, depending on their main function, as follows:

(i) 'Pre-trip information services', ie services which help drivers plan their routes and departure times;

(ii) 'Roadside displays', ie roadsigns which help drivers follow their route plans over the ground;

(iii) 'Traffic information broadcasting systems', ie broadcasting systems which inform drivers of changing road conditions and the availability of routes while they are driving;

(iv) 'Electronic navigation aids', which can be further divided to distinguish:

'Vehicle based electronic navigation aids', ie navigation aids which are self contained within the vehicle, and

'Network based electronic navigation aids', ie navigation aids which depend upon an infrastructure of roadside equipment. (2)

Pre-Trip Information Services

Drivers generally plan their journeys using maps - which may be printed on paper, or 'held' in the driver's memory. Maps however show only which roads are available. Supplementary information is required to describe road and traffic conditions, and to show which route is 'best' at a particular time.

This supplementary information is provided by the 'pre-trip information services' which include the media, (ie newspapers, pamphlets, radio and TV) and the telephone.

Newspapers and pamphlets are particularly well suited to giving drivers a broad picture of predictable events and are widely used, for example, in both France and the UK to advise drivers of preferred routes and departure times during peak holiday periods.

Traffic information is also available during the normal news bulletins broadcast by radio and TV as well as the Teletext services, eg CEEFAX and ORACLE in the UK, which are provided by the broadcasting companies. In most developed western countries several pages of Teletext information are devoted to describing conditions on major roads, and in some cases problem areas are indicated on a stylised map display. In the ANTIOPE service in

France, maps showing the major roads in selected areas are updated hourly by the Traffic Police so that drivers can readily see which roads are congested and which are not, and plan their route accordingly.

A more recent development involves the use of Videotex, which unlike Teletext allows the enquirer to interact with the system in order to obtain the answers to specific questions, and is demonstrated in the experimental ROUTE-TEL system developed by TRRL.

In this a model of the road network, currently comprising some 10,000 nodes from the Present Year Network File (PYNF of the Department of Transport, is contained on a central computer together with algorithms which enable best (ie, minimum time, distance or cost, ie a trade off between time and distance) routes to be calculated in real time. A driver then gains access to these programs via a remote Videotex terminal and is presented with a response page which asks him to specify the origin and destination of his proposed journey, his journey criteria (eg, minimum time, distance or cost), and his vehicle type (eg, car, car and caravan, heavy goods vehicle). In return he receives a list of route following instructions including warnings of any hazards, as shown in Figure 1. The origin and destination can be specified in terms of node numbers, grid references, post codes, or place names.

- - Route - Tel - -	Quickest Route
From: Brighton	
To: Guildford	
Go North on the A23	
Left onto the B2155	at 15.0 miles
Left onto the A279	at 17.8 miles
Right onto the A281	at 19.0 miles
Straight onto the A248	at 40.3 miles
Straight onto the A281	at 40.4 miles
End at turnoff for A3	
Estimated journey dist.	42.1 miles
Estimated journey time	68 minutes

FIGURE 1 Computer Generated Route Plan from ROUTE-TEL

A possible future development might lie in giving the motorist direct access to such computer based route planning services - perhaps via Videotex (eg. British Telecom's PRESTEL)

There are around 40,000 significant junctions in Great Britain so that a comprehensive route planning service would have to allow for up to $40{,}000^2$, ie 1.6×10^9 possible journeys. Many of these would be trivial and could probably be ignored, but with current PRESTEL charges running at about £1 per page of memory per year, the annual cost of permanently storing a comprehensive set of routes would be prohibitive. A much cheaper solution would be provided by a dedicated computer which could compute optimum routes in real time and which could in turn be accessed by the PRESTEL computer via a gateway.

Making a number of assumptions about the likely cost of such a system, eg, cost of a dedicated computer of necessary size, accommodation, staff required, running costs, British Telecom equipment connection and hire charges, etc, it is estimated that the amortized cost of setting up and running a Videotex route planning service would be £0.7M per year for the first ten years. It is estimated that a typical call might cost aout 30 pence to a 'typical' user who requires route plans for 50 trips each year.

For 30p however the motorist gets only a route listing displayed on a TV screen. Alternatively, the instructions could be provided in the form of a 'route card' which could be incrementally stepped through a special in-car display unit. Such a device has been demonstrated at TRRL where the route card is in the form of a digitally encoded magnetic tape cassette. The in-car unit needed to read the cassette incorporates a micro-processor unit, a cassette reader, a liquid crystal display unit, and a speech synthesis unit. Route following instructions are then both spoken and visually dislayed to the driver who must press a button to obtain the next instruction after the last has been complied with. It is estimated that such a device could be marketed for around £60.

An alternative method of presenting route information has been demonstrated by Robb & Morrison of Glasgow University who have used a computer to generate route maps directly. The maps can be individually tailored to meet the driver's journey requirements, eg, to show the location of garages and service areas en route. The maps are generated on a mainframe computer, but are accessed from a BBC micro-computer which acts as an intelligent terminal, and is used to locally control the scale and detail of the information displayed on a VDU, and then to print it on a printer.

Roadside Displays

A driver who has planned his route must generally rely on roadsigns to guide him over the ground. Road signs may be fixed or variable in the information they display.

Fixed Roadsigns. Roadsigns on trunk roads in England and Wales are the responsbility of the Department of Transport. Signing on other roads is the responsibility of the County Councils. Guidelines for signing are given in the Department of Transport's Traffic Signs Manual, but what destinations appear where is often a matter for the judgement of an officer who is responsible at the time. UK Transportation Consultants Wootton Jeffreys & Partners (1981), have suggested that computing techniques may provide an improved method for deciding what placenames should appear on direction signs. They suggest that placenames should be hierarchically ranked in terms of relative traffic importance, and that this ranking can be used to indicate a 'distance limit'. Their SIGNPOST programs can then be used to trace out 'best' routes from each placename to all junctions that fall within a route distance equal to the 'distance limit'. The information on the signs will then be rational, ie, designed according to fixed criteria, and consistent, ie, once a placename has been encountered on a sign it should appear on all subsequently encountered signs until the place itself is reached.

Variable Message Signs Variable road signs, or Variable Message Signs (VMS - as they are more commonly known) are used extensively in many countries for route control and guidance purposes - mostly at known bottlenecks such as tunnels and bridges, where a limited number of clearly identifiable and local diversion routes are available, and on motorways to divert traffic off when the way ahead is closed. They are also used to convey a multiplicity of other information.

VMS on UK Motorways. VMS are used on motorways in the UK to give drivers warnings of an unspecified hazard on the road ahead. The signs can be used, for example, to indicate reduced speed limits, the number and position of lanes which are closed to traffic, or to show 'stop' or 'all clear' messages. The newer design of signs (or 'signals') are of the fibre optic type in which a single bulb is used to illuminate a bundle of optic fibres whose other ends are arrayed in a matrix to form a particular sign or symbol. A second bulb can be used to illuminate a second bundle of optic fibres, and so on. Up to 16 separate messages are possible in one sign.

The motorway system in England and Wales is organised from 43 police control centres and the system for controlling the signals is currently being upgraded so that each of these centres will have its own computer for managing its own signals. In this system, each control centre is connected via a Regional Communication Controller to a communication link which is configured to form 4 closed loops which together cover the whole of the motorway network. No central controller is used; each 'message' entering the loop is passed from regional communication controller to regional communication controller around a loop (and if necessary, between loops), until its destination is reached. Thus if a loop is broken the message is automatically re-routed around the next preferred loop. Signal control messages are digitally encoded, and the operator receives confirmation when the messages reach their destination and the signals are set. The control computer also checks that adjacent signals are set consistently and do not, for example, show widely different speed restrictions. A self test facility automatically checks the signals at regular intervals and reports any faults.

Other uses of VMS. VMS can be used to provide both route control and ramp metering functions at motorway on-ramps. The systems employed in Glasgow's CITRAC traffic control scheme - see for example Mowatt and Young, (1984) use fibre optic technology to show speed limits, together with 'secret signs', (ie, signs which are normally blank, but show a message or symbol when switched on), and switchable direction signs to show when a ramp is closed and a diversion is in force. When the on-ramp is subject to a speed restriction, this is shown on the motorway signal; when the ramp is closed, flashing lights on an advance switchable direction sign indicate that motorists must follow a signposted diversion.

Similar VMS systems are employed in the USA, Japan, France and Germany for ramp metering and route control, though they use different types and configurations of signs. In Italy, Teletext is used to provide drivers with information on road conditions at the entrance to toll booths on some motorways.

VMS can also be used to show lane availability in 'Tidal Flow' traffic schemes; matrix signs, mounted on gantries to provide one sign over each lane, can show lane availability, by either a white downward pointing arrow or a red X. One example of such a scheme is on London's Albert Bridge.

Variable message Signs can be used in integrated car parking schemes to show which of several car parks have spaces and which are full (examples from the UK include Glasgow and Torbay). These signs are usually supplemented by fixed signs to guide drivers to the car parks.

Secret signs are sometimes used at particular sites in both the UK and abroad, to show the drivers of heavy or high vehicles that they have tripped a sensor and that they cannot use a bridge; and to show drivers when they are exceeding a speed limit or are following too close. They have also been tried in Germany to give specific warnings at sites experiencing queues and fog.

Research on improved and enhanced VMS is currently in hand in most developed countries, and recommendations for a standard set of messages and symbols for use in Europe have recently been produced under the auspices of COST 30 bis (1985).

The costs of a VMS installation can be expected to vary widely, depending on the number and type of signs used, and the communication infrastructure required for controlling them remotely. Typical costs fall between about £3,000 per sign for a simple installation and £15,000 per sign for a motorway signing system. (3)

Traffic Information Broadcasting Systems

Special arrangements for broadcasting traffic information over radio exist in most countries. The systems are used mainly to advise drivers of traffic incidents and hazards, and to encourage them to divert or postpone their journeys in order to alleviate congestion and delays.

Organisation and facilities in the UK. There are some 78 local radio stations in Great Britain. They serve the main centres of population and have developed their own arrangements for broadcasting traffic information with the local police force and the motoring organisations.

About 80 per cent of cars in the UK are now fitted with radios. Most of these can receive medium and long wave transmissions, but only about 50 percent are capable of receiving on VHF. This is in marked contrast to some continental European countries where car radios are predominantly VHF.

Local radio stations broadcast between 2 and 20 bulletins per day at regular times depending upon demand.

In addition, all police forces have access to the BBC Motoring Unit who arrange for traffic incident and predictable event information to be broadcast over the BBC national network.

The Unit is manned 24 hours per day and is responsible for assessing the importance of the information from police forces and deciding whether to initiate a broadcast via programme control staff operating the BBC national network. Criteria for message content and justification have been agreed with police. They rest mainly on whether the road is a principal road or a motorway and, if it is, whether it will be blocked for more than one hour, and on the reason for the blockage. A further requirement is that any diversion instructions must be simple. These criteria, together with the urgency of the information and availability of 'air time', are the principal factors which decide whether a broadcast takes place.

Information on predictable events is also received by the Unit from local authorities, public utilities and the motoring organisations, but it is checked with the appropriate police force before a broadcast is requested. Predictable information is usually broadcast at regular times on 3 of the 4 BBC national stations, but information on emergency incidents is usually transmitted as 'flash' interruptions to programmes. Motoring 'flashes' initiated by the BBC Motoring Unit averaged 500 per month in 1976, with a peak of about 900 in December.

Drivers need no special equipment in order to receive these broadcasts.

The German ARI System. Broadly similar broadcasting procedures are used in most West European countries, although parts of West Germany, Austria, Switzerland and Luxembourg have adopted the Blaupunkt ARI (Autofahrer Rundfunk Information) system, which has some special features.

The main function of the ARI system is to assist in tuning to the broadcasting stations which provide traffic information, and to alert the driver when a traffic broadcast is imminent. In West Germany some 65 per cent of cars are fitted with VHF receivers, and traffic information is broadcast over a network of about 40 stations, each transmitting on its own frequency in the VHF band. Because of the characteristics of radiation in this band, the range of each station is approximately line-of-sight. The country is divided into about 12 traffic information zones, each served by several of these stations. With such a system using a large number of short range stations each radiating on its own frequency, it is essential to provide some tuning aid to select the frequency of the station covering the area in which the vehicle happens to be travelling.

Each transmitter equipped with ARI radiates a special sub-carrier of 57 kHZ. This sub-carrier is modulated with one of several low frequency tones which identifies the stations which are serving the zone in which the vehicle is travelling, and is modulated by a further tone whenever a traffic announcement is taking place. To utilise these special signals the vehicle must be fitted with a decoder associated with its VHF radio, which in its simplest form detects the sub-carrier and lights a pilot lamp when the driver tunes to an ARI station. A slightly more complex decoder also allows the driver to mute his radio (thus avoiding the necessity to listen to the rest of the programme broadcast by the ARI station) and restores the volume to a pre-set level on detecting the special modulation associated with a traffic message. A more complex decoder can be fitted, which illuminates the pilot light only when the driver tunes to an ARI station serving the zone where he is driving (he must previously select a code corresponding to the zone of travel, which he obtains from a map or a roadside sign). Another much more expensive version incorporates an additional radio frequency section which automatically tunes to the strongest ARI station serving his zone and interrupts the driver's chosen programme whenever a traffic announcement takes place. It is estimated that about 80 per cent of the car radios in West Germany have an ARI decoder fitted, though these are mostly of the simplest kind.

Several manufacturers produce car radios which are capable of decoding the ARI signals. The addition of this facility adds between about £10 to £20 to the basic price of the radio receiver.

The American HAR System. HAR (Highway Advisory Radio) is now widely used in the USA (Turnage, 1980). Two frequencies 530 KHz and 1610 KHz are allocated for this. They respectively fall just below and just above the standard AM broadcast band and can therefore be received by most AM car radio receivers in the USA. Local transmitters using one or other of these frequencies, and with a power such that reception is confined to a radius of up to about 2 km, are sited at the approaches to hazardous areas where they broadcast pre-recorded hazard warning and diversion messages from a continuous loop tape. Because of the local nature of the broadcasts, a fixed sign must be employed to tell drivers to tune to the appropriate frequency so that they can hear the broadcasts. The second frequency can be used at the far end of the hazard so that drivers approaching from the opposite direction can be given a separate set of messages.

A subsequent development of an AHAR (Advanced HAR) system (Mammano, 1984) attempts to avoid the need for manual tuning of the car radio by using a subsidiary FM receiver which automatically mutes the car radio, tunes to the AHAR frequency (45.80 MHz), and returns the radio to its initial state after the roadside message has been repeated twice. Tests on this system were tried using the transmission of actual voice messages, but the possibility of transmitting digitally encoded speech for synthetic reproduction in the vehicle was also examined.

The BBC CARFAX System. CARFAX is a proposed broadcasting system dedicated to the transmission of traffic information. The system would consist of a network of low power medium frequency transmitting stations all operating on a single channel and sited about 40 to 70km apart to cover all or selected parts of the country. Whenever a traffic broadcast was required in a particular area, the appropriate transmitter would be energised for the duration of the broadcast; interference would be avoided by prohibiting simultaneous transmissions from all other transmitters within range. This technique - known as time division multiplexing - permits nationwide coverage to be achieved with a single channel. The system would be controlled by a simple computer at a broadcasting centre which would allocate recorded messages to be broadcast by the transmitter serving the appropriate area and could give precedence to messages allocated a high priority by the traffic control authorities.

The motorist would receive the broadcast for his particular area by means of a cheap fixed tune radio receiver costing between £10 to £20. This unit could operate in conjunction with an existing car or domestic radio, or be installed as an independent unit, or form an integral part of a normal radio.

In any such system based on local broadcasts, it is desirable that the motorist is not forced to listen to the broadcasts which serve other areas, and also that he can listen to the entertainment stations of his choice or to a cassette player, while awaiting a relevant traffic message. These requirements are met in the CARFAX system by preceding each transmission by a 'start of message' code, which is detected in the special receiver. The receipt of the signal is necessary to prepare the receiving equipment for the traffic broadcast, and it can mute other programmes being received until the receipt of a 'finish' code or the elapse of a pre-set time interval.

Although the system primarily uses amplitude modulation (AM), the 'start of message' code (and the 'finish' code) is broadcast using frequency modulation (FM), and the short burst of FM is accompanied by simultaneous FM transmissions from the ring of adjacent transmitters; these simultaneous transmissions however do not include the 'start' code. This procedure is designed to take advantage of the property of FM receivers that they are 'captured' by the strongest of two or more FM transmissions radiating on the same frequency, and unlike AM transmissions can demodulate the signal from the strongest station with little interference from the others. Thus a receiver in the zone for which a broadcast is imminent would be captured by the FM transmission containing the 'start' code, and would be activated to receive the traffic message; whereas a receiver in the adjacent zone would be captured by the stronger simultaneous burst of FM from its own local transmitter, but as this does not include the 'start' code, the receiver would not be activated and the subsequent message would not be received. The shape, size and degree of overlap of these zones could, to some extent, be tailored to meet the requirements of the traffic authorities by adjustment of the transmitter output powers.

The system is fully described in Sandell and Edwardson (1977).

<u>Cellular Radio</u>. The future of CARFAX in the UK is uncertain, mainly because of problems in allocating suitable frequencies. But

the cellular structure of the ring system is almost identically repeated in the infrastructure required by a Cellular Radio system. Two such systems are now operating commercially in the UK and there is a possibility that the principles of CARFAX could be achieved using Cellular frequencies.

The Cellular radio system adopted in the UK is based on the Bell System Advanced Mobile Phone Service (amps) and is designed to provide a portable, mobile and fixed station, radio telephone service using a 'cellular' radio coverage plan. The system operates in the frequency bands between 890 MHz to 915 MHz and 935 MHz to 960 MHz.

In the system, the coverage area is divided into a number of abutting cells whose radii vary from about 16km in rural areas down to about 2km in the most busy urban areas. Each operating company has about 220 duplex radio channels, each channel having a band width of 25KHz. It is not possible to use all the radio channels in each cell as the co-channel interference from adjacent cells operating the same frequency would be intolerable. The cells are therefore arranged in clusters of seven and the available channels are divided equally among them. These seven cell clusters are then repeated over the country so that the frequencies are reused, but from the geometry of the cell clusters, co-channel interference can only occur from cells located some distance away. The interference effects are thus minimised.

At the heart of the cellular network there are a number of computer controlled mobile-radio switching centres (MSC), each controlling a subset of the base stations. The MSCs are similar to public telephone switches, but they have additional hardware and software for operation within the cellular network and for interconnection with the PSTN, (Public Switched Telephone Network).

One of the channels allocated to each base station, ie. cell, is used as a control channel. This carries digital information and is monitored by the MSC to determine signal strength, and on the basis of this to allocate particular mobiles to particular base stations. The MSC also uses this channel to page a mobile to whom a call is being made; and to regularly interrogate mobiles to determine when they wish to make a call. Calls are thus set-up by the MSCs which allocate channels as appropriate from amongst those available to the base stations.

As a mobile caller moves across a cell boundary, a 'hands-off' procedure occurs in which the radio channels are automatically changed so that the mobile can be served by the base station for the new cell.

In essence, cellular radio enables the public telephone network to be extended into vehicles, and anything which can be achieved on the public network can, in principle, be achieved on the cellular system.

Cellular radio therefore offers potential with regard to the Videotex systems described previously. A cellular radio could readily be used in combination with a portable micro-computer and a modem to enable a driver to interrogate a remote computer directly from his vehicle cab. He could thus have immediate access to ROUTE-TEL for route plans, or to a data base which could provide updated versions of the electronic maps required by the CARIN, EVA and NAVIGATOR types of navigation aids, which are described in the next section.

Like domestic telephones, mobile cellular 'phones are usually leased from the operating company who in addition levy a time dependent charge on each call.

The Radio Data System. The radio data system (RDS) of the European Broadcasting Union (1984) enables digitally encoded data to be superimposed on the stereo multiplex signal of a conventional FM radio broadcast. This data can then be decoded by a suitably adapted car radio, which can, in turn, be made to automatically 'hop' to the stronger of two signals when a driver leaves the reception area of one transmitter and enters that of another which is broadcasting the same programme.

There is some spare capacity on this data channel and discussions are in hand internationally to consider whether traffic information could be included with the programme data - see for example COST 30 bis (1985). A special vehicle unit would be required to decode the traffic information which could then be displayed either as internationally recognisable symbols, or as synthetic speech. Such a system is particularly attractive from an international point of view because digital transmissions are independent of language barriers and messages could be reconstructed as synthetic speech in the language of a driver's choice.

Electronic Navigation Aids

Various electronic aids have been developed to help drivers plan and follow routes. Some are self contained and can be wholly carried in a vehicle, while others depend on an infrastructure of equipment at the roadside. But all rely on a navigation aid to enable the vehicle to keep track of its position in the network.

Navigation Techniques. Navigation implies knowledge of position. There are basically three techniques for navigation (i) dead reckoning; (ii) trilateration; (iii) beacons - any of which either singly or in combination, may be used to update a vehicle's position.

Dead Reckoning. Navigation aids which use a dead reckoning technique rely on distance and heading sensors fitted to the vehicle so that progress from a known start position can be continually monitored and position, in terms of grid references, updated.

Dead Reckoning devices have been manufactured for military use for many years but only recently with the advent of cheap micro computers has it become economically worthwhile to adapt the technology for use in private vehicles.

The distance sensor may be a mechanical take off from the odometer, or more usually some mechanism for sensing wheel revolutions, eg. a series of magnets equally spaced around the wheel rim or drive shaft and sensed by a magnetically activated switch fixed to the body of the vehicle. A pulsed or digital output is clearly preferred for detection by a microcomputer. Accuracies of about 2 per cent of distance travelled can be obtained. Higher accuracy is precluded by variations in wheel diameter which occur as the tyre wears, or inflation pressures vary, but can be improved by regular calibration.

The heading may be deduced by measuring steering wheel rotation, or by monitoring the differential rotation of the rear wheels of the vehicle. More generally however, a special purpose sensor is used, such as a gyro compass, or an 'electronic' compass, eg. a flux gate sensor. These last are usually an energised pair of coils wound at right angles. The resulting field, which is affected by the earth's magnetic field, is then measured to determine the orientation of the gate (and hence the vehicle) with respect to the earth's field. Careful positioning of magnetic heading sensors is required as they are generally affected by the electric circuits and the mass of metal in the vehicle itself. As a result they must usually be calibrated for each individual installation. However they are also affected by metal masses, eg. bridge structures, other vehicles etc, outside the vehicle so that high accuracies are precluded.

Dead Reckoning systems thus accumulate error both in distance and orientation. As a result they need to be re-initialised typically after a few hours of driving, or after a few tens of kilometres have been driven.

Trilateration. Navigation aids which use a trilateration technique rely on the detection of radio transmissions from three or more fixed points. Decca and Loran-C are well known examples for use at sea and in the air. On land however the signals are confused by multi-path reflections from tall buildings and hills, so that position accuracies of better than about 200m are difficult to obtain.

The satellite navigation systems such as SATNAV and GPS (Global Positioning System) are also based on trilateration - using transmissions from satellites in (mathematically speaking) well behaved orbits. Again these systems are fine for sea and air use, but are confused on land, and particularly in urban areas, where tall buildings may prevent enough satellites from being within line of sight to enable a fix to be made. Accuracies of better than 200m are therefore difficult unless some interpolation of position can be made between fixes.

Beacons. In these systems beacons which may be passive or active, are positioned at known points in the network. A passing vehicle is then able to update its position by interrogating the beacons. In practice these systems are rarely used in isolation because it is not usually economic to deploy sufficient beacons to provide very high location accuracy. A Dead Reckoning technique is therefore frequently used to interpolate between beacons. Such systems are commonly used by buses which follow fixed routes - in these cases distance information only is required for interpolation.

The next two sections show how the 3 basic techniques of navigation may be incorporated into electronic aids for route planning and following.

Vehicle Based Electronic Navigation Aids

In this section we consider the wholly vehicle contained systems. There are three main types: (i) simple directional aids; (ii) map displays; and (iii) route guidance aids.

Simple Directional Aids. These comprise a heading and a distance sensor, a microcomputer, a numeric keypad, and a display unit. The driver enters the grid references for his current position and required destination on the key pad. The micro computer then computes the vector connecting these positions and displays it in terms of a 'crow flight' distance and a heading, on the display unit. Heading is usually displayed as an arrow which physically identifies for the driver the direction he must take in order to reach his destination.

As the vehicle moves off on the journey, the microcomputer continually monitors the heading and distance-travelled sensors in order to recompute the vehicle's current position, and to update both the remaining 'crow flight' distance and heading displays.

Thus as the driver approaches each junction on his journey, he has available both a measure of how near he is to his required destination and the direction he should take to reach it.

Examples of such systems are the NAVICOM from Toyota, and DRIVEGUIDE from Nissan. A similar device CITYPILOT from VDO of West Germany uses a specially prepared set of maps together with a lightpen connected to the vehicle unit to facilitate the driver entering the grid references for his origin and destination.

These devices can be expected to work well in regular (ie. matrix) networks, but fall down in irregular networks such as in the UK where roads do not always continue in the same direction as that from which they leave a junction. Nor can they help to identify an alternative route which may be faster when this requires the driver to head off in altogether the wrong direction at some point in order to pick up, eg. a motorway.

Map Displays. An alternative, and more usable method of presenting navigation information, is to superimpose the vehicle's position on a map of the surrounding road network. Such techniques are not new. They have been used by the military for many years. But only with the advent of cheap micro computers has it become possible to adapt the technology for use in private vehicles.

Probably the first successful attempt was made by Honda with their Electro-Gyrocator. This device relies on a gas rate gyro to sense heading and on a magnet type distance sensor to dead reckon the position of the vehicle at successive increments in time in the same way as described above. The output however is a moving spot of light on a CRT. A suitable map foil is then placed over the screen and the position of the spot of light manually adjusted to correspond with the vehicle position. Other knobs and switches allow the driver to calibrate the system to correspond with the scale of the map foil. The driver then marks his destination on the map foil and sets off on his journey. After he has travelled a kilometer or so he (or his passenger) must align the map to correspond with the trace of the spot of light. Thereafter his route thus far and his present position on the map are clearly shown. A subsidiary spot of light shows the orientation of his vehicle on the map. He can thus see where he is in relation to his destination, and which direction he must take in order to reach it.

Later developments such as ETAK from USA and CARIN from Philips in Holland have replaced the map foil with a computer generated map which appears on the CRT together with the spot of light which shows the vehicle's position. In these systems extra computer memory is required in which to store the map information in digitised form. ETAK uses cassettes, while CARIN uses a CD ROM (ie. compact disk) which is capable of storing around 600 Mbytes of data - several Mbytes are required for a digitised map covering a large town.

For the future Philips propose that CARIN should use GPS. In situations where an accurate position fix can be made, this will save the driver from having to initialise the system. It should also improve the accuracy of position updates when used in conjunction with the Dead Reckoning system, but it is unlikely that it can substitute for this.

Route Guidance Aids. The simple directional aids and the map display devices described above tell the driver about his position relative to his destination, but cannot offer advice on which is the 'best' route to follow. To do this a more comprehensive description of the network must be stored in the vehicle, together with an algorithm which can operate on these data to compute a minimum path through the network.

In some devices, such as the Mercedes Benz ROUTEN-RECHNER, Blaupunkt's EVA, and ROESY from Wootton Jeffreys & Partners in the UK, a description of the road network is stored on a suitable memory device in terms of the intersections (nodes) and the impedance of the sections of road (links) which connect them. A suitable algorithm, eg, D'Esopo's, can then be used to compute the minimum impedance path between any two intersections. Distance is most commonly used as a measure of impedance because it is the easiest to establish. But drivers generally prefer time, or cost (ie, a trade off between time and distance) as a criterion for route selection, and would sometimes like 'most scenic'.

In other devices, eg, the TRRL NAVIGATOR, the network description is pre-compiled to provide a 'Signpost' for each junction with the resolution needed to give guidance to every other junction in the same network. The pre-compilation process can employ a distance, time, or cost criteria - depending on the purchaser's preference. In the CARGUIDE system from Carnegie-Mellon University, a hybrid system is used; the street network is divided up into 'zones', routes between zones are then pre-compiled, but routes within zones are computed on demand.

In all cases the driver must initialise the system by keying in codes for his start position and the intersection which most nearly corresponds with his required destination. The device then computes the 'best' route through the network and gives the driver his first instruction. This may be via one, or a combination of, alpha-numeric displays, speech synthesis units, or graphical display units.

ROESY was perhaps the first system demonstrated in a vehicle, based on an Apple II micro computer with 48K of memory and a 5.25" floppy disk drive. Special Programs enabled the user to specify his own network of up to about 300 nodes and 450 links, and to specify both the lengths and impedances of the links. Origin and Destination codes were entered on the Apple keyboard, and a speech recognition package was tried. The system used a speech synthesis unit to speak 'turn left' etc, instructions to the driver as he progressed on his journey. It was assumed that the driver would not deviate from the planned route - a measure of distance travelled was then sufficient to 'locate' the vehicle and to indicate when successive instructions should be given. The device was not developed beyond the demonstration phase.

ROUTEN-RECHNER from Daimler-Benz in Germany is similar. It gives guidance on the German motorway network using a minimum distance criterion for route selection, and instructions are presented to the driver on a small alpha-numeric display unit. Again it is assumed that the driver follows the instructions implicitly - a measurement of distance alone thus enables the system to know when to display successive instructions.

The TRRL NAVIGATOR is similar, but uses a visual display comprising directional arrows and alphanumerics, as well as a speech unit to instruct drivers, for example, to 'turn left on the A420'. The journey criterion can be distance, time, or any combination of the two (ie, cost), and the network is based on the 10,000 node and 15,000 link PYNF network of England owned by the UK Department of Transport. Again it is assumed that the driver does not deviate from the planned route so that location can be deduced from a measurement of distance alone.

EVA from Blaupunkt in Germany employs both distance and heading sensors in order to detect when the driver has deviated from the planned route, to automatically re-initialise the system, and to work out a new best route from the 'new' position. EVA also employs a speech synthesis unit to give 'turn left' etc instructions as well as a Liquid Crystal display which shows the junction layout with the required route through the junction superimposed.

CARGUIDE is based on a map display type of device. The driver enters his Origin and Destination in terms of intersections, using two street names to identify each. He receives instructions as sythesised speech 'straight, left, or right' and the name of the street to take. A visual display shows the surrounding road network with the current and next links flashing. The driver must press a button each time he obeys an instruction in order to get the next.

The CARIN system described in the previous section has also been developed to provide similar route guidance facilities but without the need to press a button.

The costs of vehicle based devices can be expected to range from about £100 in the case of the simple directional aids, through a few hundred pounds in the case of NAVIGATOR and ROUTEN-RECHNER, to more than £1,000 in the case of ETAK and CARIN which incorporate sophisticated navigation systems and graphical displays. Mass production and the use of VLSI (Very Large Scale Integration) techniques may reduce these costs in the future.

Network Based electronic Navigation Aids

In this section we consider the electronic aids to route planning and following which rely on an infrastructure of roadside equipment to provide location and route guidance information. These are essentially automatic route guidance systems with the potential to give hazard warning and guidance advice in real time.

In practice a wide range of systems is possible and COST 30 bis (1985) shows that the total costs of systems are strongly influenced by the complexity of the roadside infrastructure required. This is to be expected because there are many times more vehicles than road junctions. Systems which can distribute most of their intelligence at the roadside will therefore involve high costs to central funds, but will require relatively cheap vehicle units.

The total costs of the system will then be minimised. In contrast, systems which require most of their intelligence to be carried in vehicles will involve more expensive vehicle units. The complexity and costs of the roadside infrastructure will then be reduced. But the total system costs will be increased.

Work undertaken by the TRRL and reported by Jeffery (1981) suggested that for economic viability some investment from central funds would be required to identify junctions by means of roadside equipment. The work then progressed to fully exploit recent developments in the techniques of IT in order to define a range of preferred systems. Three of these are described below.

A Junction Beacon System of Automatic Route Guidance. A practical automatic route guidance system which requires the lowest contribution from central funds must utilise the cheapest means for uniquely identifying each junction to an approaching vehicle. A reliable scheme would be achieved by a system of junction beacons; ie. the beacon system of navigation described previously, which employed either a passive technique, such as a coded pattern of magnets buried in the road surface, or an active technique, such as the transmission of a suitable code over a one-way optical, ultrasonic, radio frequency, or microwave communication link. The code could then be detected by an approaching vehicle, and the junction as well as the vehicle's position, uniquely identified.

For reasons of practicality and cost it was thought that a one-way radio frequency communication link should provide the best method. A radio beacon could be fixed to, and powered from, any convenient lamp post, operated at a frequency of say 450 MHz and modulated to carry a frequency shift keyed digital 'junction identifier' code at 64K bits per second. The transmission could then be readily received by a relatively cheap and compact car-borne radio receiver unit. At most junctions a single 'long range' transmitter could be sited at the centre of the junction. With the aerial mounted at a height of 10m, an effective radiated power (ERP) of about 2.6W should be sufficient to enable vehicles to receive the transmissions at a range of greater than 500m. At around one in four junctions however, proximity or topographcial effects may cause vehicles to receive the wrong signal or to receive a signal too late. At these junctions 'short range' transmitters would be required to idontify each junction approach road individually and at a few, particularly urban sites, several junctions may need to be combined and treated as a single compound junction. These transmitters would need to be sited about 500m before the junction, again perhaps on a suitable lamp post. With the aerial located about 7.5m above the carriageway an ERP of several microwatts would be required.

A vehicle would need to be equipped to receive and decode the beacon transmissions. It would also need a vehicle unit comprising a keypad, a microprocessor, an alphanumeric display unit, and an 'electronic map', ie. one of the map display or route guidance aid type devices (such as CARIN, EVA, or NAVIGATOR) described previously. But in this system no navigation subsystem would be needed in the car, and the driver would not need to specify his start position. He would simply enter the code for the junction nearest his destination on to his keypad and set off on his journey. As he approached each equipped junction in turn his in-car equipment would receive the beacon signal and decode it. The microprocessor would then search the 'electronic map' which could, as in the TRRL NAVIGATOR system, take the form of a table of information associated wih each junction. Once the table for the particular junction had been located a second search exercise could identify the particular entry in the table which corresponded with the driver's destination code. This entry could in turn identify an appropriate instruction which could then be simply displayed to the driver as a directional arrow, together with a road number, and where necessary, a compass direction. The driver would thus be informed of what signed path he should follow through the junction in order to reach his destination by an optimum route.

The costs to central funds for such a scheme are estimated to be about £2000 per typical junction for the provision and installation of radio beacons, plus, ideally, about £420 to provide extra road signs on each exit road so that the driver could confirm he had identified the correct exit. The total contribution required to equip 40,000 significant junctions in Great Britain would therefore amount to about £100M. The likely cost of equipping an individual vehicle might amount to around £300. If one in four, ie, four million or so vehicles in Great Britain were to be equipped the total system costs would therefore sum to about £1300M.

An Electronic Signpost System of Automatic Route Guidance Using Radio Beacons. In the previous system each of up to around 20 million vehicles in the UK are required to carry their own electronic map, each of which in turn duplicates the coded direction table for each of the 40,000 significant junctions in Great Britain. This technique is clearly wasteful of data storage facilities and a considerable saving might be achieved if that part of a single electronic map which was relevant to a particular junction could be broadcast directly from the junction itself. A single electronic map could thus be distributed to the roadside where it could be shared by all drivers. As each vehicle approached a junction it could then monitor the broadcast, which would in effect provide an extremely high resolution electronic signpost.

The junction beacon scheme described in the previous Section was designed to have sufficient spare capacity to enable the entire coded direction table for each junction to be continuously transmitted by the junction beacons and received up to five times by an approaching vehicle. The modifications required are therefore minimal and it is estimated that while the cost of equipping a typical junction would increase to about £2800, the cost of equipping a vehicle would decrease to about £160. The total system costs for four million vehicles and 40,000 junctions would therefore amount to about £750m of which £110m would be required from central funds.

The total system costs are therefore reduced by around 40 per cent at the expense of only a ten percent or so increase in the cost to central funds. Moreover, a further advantage of an electronic signpost system occurs because the electronic maps are contained within roadside units. The maps must therefore become the responsibility of a central authority who could alter them to take account of changes in the network which may be incurred due to new road building, or long term diversions, etc. The system could therefore be made adaptive because all equipped vehicles would instantly recognise and follow any new route instructions as soon as they were given.

Adaptability could be achieved at two levels: either slowly and relatively cheaply, by manually changing the memory units containing the map information, or more rapidly and at considerable extra cost, by remotely controlling the map memory contents directly from a traffic control centre.

In the first case only predictable and long term diversions could be catered for, but the second case could in principle allow a more comprehensive system which could respond quickly, and in real time, to divert traffic around incidents as they occur and perhaps to give drivers warning of hazards on the road ahead.

A limitation of the system occurs however because only a single radio beacon is required at most junctions. A vehicle using the system cannot therefore always identify which particular approach road it is using and ideally, subsidiary roadsigns are required so that drivers can identify the individual exits. The guidance information which can be given to the driver is therefore restricted to a simple instruction showing which roadsign should be followed.

Since drivers are familiar with the existing roadsign system, this arrangement should prove satisfactory in practice.

Nevertheless, if each approach road could be uniquely identifed by its own radio beacon it would be possible to give more versatile guidance instructions. For example, a plan of the junction layout could be displayed together with the driver's required route superimposed and an indication of which lane the driver should adopt on both entering and leaving the junction. In addition, the beacons on the individual exits could be used to confirm that the driver had taken the correct exit and to warn him if he had not.

In such a scheme no modification or provision of roadsigns would be required but the cost of equipping a typical junction would nevertheless increase substantially to about £4800. At the same time a more complex display would be required so that the cost of a vehicle unit might rise to about £180. The total system costs for four million vehicles and 40,000 junctions would therefore sum to around £900m of which about £190m would be required from central funds. In practice the additional benefits offered by this scheme are probably insufficient to outweight the extra costs; and where a more versatile display is required (perhaps to guide drivers in busy urban areas), it could be obtained at substantially less cost (as shown below) if the one-way "road to vehicle" radio communication link were replaced by a two-way inductive loop communication link.

An Electronic Signpost System of Automatic Route Guidance Using Buried Loops. Radio transmitters are relatively cheap to buy and install, and they provide a system of junction beacons which involves a relatively low investment from central funds. However, problems of signal interference dictate that the vehicle unit must incorporate considerable and costly error detection and correction logic in order to ensure that the beacon signals are reliably received. The cost of the vehicle units is therefore to some extent inflated in order to keep the roadside equipment cost as low as possible.

A more satisfactory means of communicating between a roadside and a vehicle unit might be provided by inductive coupling which would rely on buried loops. The position of these relative to vehicles passing over them could be very well defined, and although such high data rates could not be achieved (around 10,000 bits/sec would be possible) the integrity of data transfer would be much improved.

The lower data transfer rate, and the physical dimensions of the loops which must be buried in the road surface, would generally prohibit the possibility of a one-way communication link as used in the radio beacon systems. A two-way link would therefore be required in order that the vehicle could communicate its destination to the roadside unit which could then look up the coded destination table and reply to the vehicle before it had driven off the loop.

The savings which could be achieved by the need for less error detection and correction logic would therefore be partly eroded by the need to equip each vehicle with a transmitter, but the cost of a vehicle unit is nevertheless estimated to reduce to about £140 for a 'simple' display, or £160 for a complex display which is capable of showing junction layouts.

At the same time the cost of the roadside equipment would increase. A buried loop could not broadcast information over a wide area and it would therefore be necessary to provide loops for each approach road separately. In addition a receiver would be required in order that the roadside units could interrogate vehicles for their destination codes. The costs of equipping a typical junction are estimated to be about £4600.

The total system costs for four million vehicles and 40,000 junctions would therefore sum to around £730M for a simple display, or £820M for a complex display, of which around £190M would be required from central funds in either case. This system offers minimum cost to both the community as a whole and the individual motorist, but requires the highest contribution from central funds. The system is the most cost beneficial of all the automatic schemes considered.

Moreover, the basic equipment costs are such that it would probably be reasonable to anticipate that a fair proportion of motorists would be prepared to buy the necessary vehicle equipment -

although a market research exercise would be necessary to confirm this.

Other Automatic Route Guidance Systems. The automatic route guidance systems considered above are essentially historic systems, and further benefits may be available to systems which can respond to traffic conditions in real time so as to help reduce congestion and delays.

In early UK and US proposals, and in the current German ALI (Bragas, 1979) and Japanese CACS (Yumoto et al,1979) demonstration systems, a two-way communication link between the roadside and the vehicle is achieved using buried loops which double as vehicle detectors. The roadside units are then controlled in real time by a central traffic control computer which monitors the flow of traffic passing over the loops, uses this information to deduce which roads are 'free' and which congested, and then controls the guidance information transmitted by the roadside units so that individual vehicles are diverted over routes which are optimum at the time.

The control computer can also instruct the roadside units to transmit hazard warning messages which can be displayed on the vehicle units when a hazard, such as fog or a traffic accident exists on the road ahead. In addition, the two-way communication link enables specific vehicles to be readily located. In some schemes, emergency service vehicles are made to transmit an identification code along with their destination code so that they can be remotely identified as they pass each roadside unit in turn, and a further relatively simple development would enable the system to give them preference by diverting other vehicles out of their way. Systems organised in this way can also be easily extended to distinguish between different classes of vehicle so that, for example, lorries can be advised against using particular parts of the network. It would then be a relatively simple step to initiate a road pricing scheme.

An automatic route guidance scheme using buried loops and a two-way communication link would therefore seem to offer not only the cheapest automatic system from both the community and individual motorist's point of view, but also the system with the greatest potential for further development at lowest cost. (A system using a one-way radio communication link could clearly not be developed so cheaply to the same extent.)

The costs of realising a fully responsive guidance and control system would not however be low. The basis of the system could be provided by the automatic route guidance scheme described in the previous section, but substantial further costs would be incurred. Nevertheless the additional expense might well be warranted by the extra benefits which could flow from the extra facilities provided, such as the in-car hazard warning system, traffic restraint, automatic incident detection, reduced congestion and delays, and the provision of an automatic traffic counting, flow and speed monitoring facility.

The AUTO-SCOUT system from Siemens of West Germany is a more recent development which uses an infra rod light beam to facilitate communication between roadside units and vehicles. Data is communicated at a rate of 64K bits/sec and Siemens propose that only 1 in 5 or so of junctions need to be equipped. At these junctions the data transmitted to vehicles provides a description of the local road network and the best routes to the next equipped junctions. This clearly saves on infrastructure costs, but at the expense of more costly vehicle units which must be able to navigate for themselves between equipped junctions. AUTO-SCOUT is therefore a hybrid between the Network Based systems discussed in this section and the Vehicle Based system described previously.

REFERENCES

Andrew C (1977). An interview survey of motorway driver requirements and signal understanding. Dept of Transp. Transport and Road Research Laboratory report LR 742, Crowthorne.

Bragas P (1979). Field testing of a route guidance and information system for drivers (ALI). Proceedings of International Symposium on Traffic and Transportation Technologies, Hamburg, June 1979.

COST 30 (1980). Final report of Theme 5: Survey of information needs. EUCO-COST 30/109/80. XII/100/80. Commission of the European Communities, Luxembourg.

COST 30 bis (1985). Electronic traffic aids on major roads - Final Report. EUR 9835. Commission of the European Communities, Luxembourg.

European Broadcasting Union (1984). Specification of the radio data system RDS for VHF/FM sound broadcasting. Document Tech. 3244-E, EBU, Brussels.

Jeffery D J (1981). Ways and means for improving driver route guidance. Dept of Transp. Transport and Road Research Laboratory report LR 1016, Crowthorne.

Mammano F J (1984). Speech synthesis for a motorist information system. Proceedings of OECD Seminar on Microelectronics for Road and Traffic Management. Tokyo, Nov 1984. 182-191.

Mowatt A M and A D Young (1984). CITRAC - the first five years. Traffic Engineering & Control, May 1984. 251-257.

Sandell R S and S M Edwardson (1977). A proposed road traffic service. European Broadcasting Union Technical Review, Dec. 1977, EBU, Brussels.

Turnage H C (1980). Highway Advisory Radio. IEEE Transactions on Vehicular Technology, 29. 183-191.

Wootton Jeffreys & Partners (1981). Effectiveness of existing signposts. A report prepared for the Department of Transport. London.

Yumoto N, H Ihara, T Tabe and M Naniwada (1979), Outline of the CACS Pilot test Systems. Paper presented at the 58th Annual Meeting of the Transportation Research Board, Washington DC. January 1979.

LIST OF PRODUCT SUPPLIERS & ADDRESSES

Blaupunkt, Blaupunkt-Werke GmbH, 3200 Hildesheim, Federal Republic of Germany.

ETAK Inc., Sunnyvale, California, USA.

Philips Research Laboratories, Eindhoven, The Netherlands.

Siemens AG, Munich, Federal Republic of Germany.

VDO Adolf Schindling AG, Schwalbach, Federal Republic of Germany.

Wootton Jeffreys & Partners, Brookwood, Woking, Surrey, UK.

EDITORS' NOTES

(1) Further discussion of this issue is contained in Chapter 3.

(2) Several alternative classifications, relating to information provided to public transport users, are suggested in Chapter 13.

(3) For further discussion on the use of VMS systems, see Chapter 6.

Information Technology Applications in Transport, pp. 353–382
P. Bonsall, M. Bell (Editors)

Chapter 15

THE ROLE OF EXPERT SYSTEMS IN TRANSPORT

Peter Bonsall and Howard Kirby
University of Leeds

INTRODUCTION

Background

Experienced judgement and specialist knowledge are essential to the proper specification, understanding and interpretation of data and computer analyses. The human expert has traditionally supplied this knowledge and judgement with the computer doing the necessary number-crunching. However, artificial intelligence (AI) research provides ways of embodying this knowledge and judgement within computer programs. Despite an early lead in the field, UK research and development into AI techniques was held back in the 1970s when the then Science Research Council took the view that the 'combinatorial explosion' of possibilities would be an insurmountable obstacle to AI development. But in America and Japan research continued, and the surge of interest in the 1980s has been a consequence of the 'Fifth Generation Computer' research programme initiated by Japan (Feigenbaum and McCorduck, 1984). This led in Europe to the ESPRIT programme of advanced technology research, and in the UK to the Alvey programme (Department of Industry, 1982). As a result, all sectors of industry have been encouraged to consider how such advanced technology can be applied, and the transport industry is no exception.

This paper sets out to explain some of the relevant techniques in simple terms, and to describe a number of situations in which transport planning and operations might be helped through their use, illustrating this by reference to the pioneering work going on in transport applications in the USA, Britain and Australia.

What the phrases mean.

Artificial intelligence implies intelligent-seeming behaviour in computers. The scope is clearly vast, including such subjects as image recognition, the understanding of natural language and the control of robots.

AI research has spawned a number of specialist languages and tools for different classes of problem. The tools most likely to be useful for transport planning and operations are known as expert systems. They involve rules and relationships devised to express human knowledge and judgemental processes. Because they involve the application of human knowledge to a problem, expert systems are often more modestly referred to as knowledge-based systems (KBS). The phrase 'knowledge-based systems' is in principle broader than that of expert systems, since it includes those situations in which the knowledge is readily available, say in a manual or a set of regulations, rather than in the experience of an expert. Another phrase is 'Intelligent Knowledge-Based Systems' (IKBS), which has become the new vogue term following its adoption in the Alvey Report (Department of Industry, 1982). IKBS was there defined to be "a system which uses inference to apply knowledge to perform a task"; it thus implies advanced kinds of expert system.

WHAT CAN EXPERT SYSTEMS DO?

Some examples.

Different kinds of knowledge-based system have evolved over the years, with differing characteristics. Some idea of the capabilities of the systems is apparent from the following list of types of problem to which they have been addressed. (The names of some of the pioneering systems are given in parentheses; these are described in most AI text-books).

- Diagnosis of disease (MYCIN); of electrical circuits (EL).
- Enumeration of molecular structure (DENDRAL).
- Detection of minerals (PROSPECTOR).
- Configuration of computer systems (XCON).

The new tools and techniques imply a fresh perspective on what is possible with computers by clients, users and programmers alike.

Client's and users' perspectives.

Knowledge acquisition and dissemination. The main advantage of an expert system for a client is that it provides a way of disseminating somebody's experienced knowledge or judgment concerning a problem to less experienced staff, or of bringing together the experience of several different experts to bear on a complex problem. This will not only make for better and speedier decision-making by the user, but also releases the expert's time for other tasks - or ensures that the expert's skills can be retained in the organisation when she/he leaves. The rules of thumb are most obviously derived directly from the expert(s), but this process can be extremely difficult and time-consuming and, in some circumstances, speedy system development can be facilitated by recording the decisions made by the expert in a given set of circumstances and then inducing the rules he is following from an analysis of his actions.

Even before the expert system itself is constructed, an advantage may be gained by the organisation if the expert's rules of thumb and judgmental values, which are usually implicit and may not at first be recognised as such by him, are made explicit by the knowledge acquisition process. Because they encapsulate expertise, expert systems also have a useful role in training less experienced staff.

Updating and maintenance. Since the knowledge of the expert(s) and the way in which it all fits together usually becomes apparent only gradually, and indeed will change over time in the light of experience and exogenous factors, an important feature of the expert system from the client's point of view is that it is easy to update the knowledge base. Knowledge-based systems are so constructed that the rules and relationships in them are transparent, and readily modified.

Treatment of uncertainty. In real life, input data is often imprecise either because of measurement error or because of fluctuations in the basic data (eg one might estimate traffic flow at 1500 vehicles/hour but it could be between 1400 and 1600 vehicles/h). Sometimes the data is more naturally described qualitatively (eg a 'high' flow) rather than quantitatively. Also, for given values of input data (no matter how precise), there may be several possible courses of action (or outcomes). Expert systems can be set up to cope with both uncertainty in input data and with uncertainty of outcome in a statistically appropriate way. (For a discussion of how this is done see below.)

User-friendliness. A prime desideratum of any computer system is that it should be easy for the client's staff to use. Clearly, this is particularly important if the intention is to make the system available to non-specialists. An expert system is not in itself user-friendly; but it can contribute to user-friendliness by ensuring that, in the machine's dialogue with the user, it asks only the relevant questions, is forgiving in its treatment of mistakes, perceives inconsistencies in the user's answers to questions, and checks for gaps in its own knowledge. It can also explain and justify its reasoning on request. A good expert system will have the user thinking it was clever enough to think of something he/she had not thought of, but recognises as true once seen.

Programmer's perspective.

Languages. Programmers familiar with conventional computing for engineering and transport applications will find that expert systems require a rather different style of approach. Traditionally they will have been concerned with the manipulation of numbers, and will probably have used the Fortran (or, more recently, Pascal) language for this purpose. These languages are fine for engineering or scientific calculations; but they are rather clumsy at manipulating rules or relationships. The AI community have developed other computer languages that are much better able to do this. One is LISP, which was originally designed (in 1958) for processing lists. Modern-day implementations, available through such products as LOOPS, KNOWLEDGE CRAFT, and KEE, are very different, in that they build powerful tools on top of LISP. They offer much more powerful and easy-to-use ways of representing knowledge, as well as good project-support environments including 'menus' and 'windows'. Another main language is PROLOG, which is designed for processing logical relationships; this is the language adopted in the Japanese research programme for a Fifth Generation Computer. POPLOG provides an environment within which codes from either of these two languages (or another - POP11) can be tied together. This environment provides editing facilities and enables small chunks of code to be edited, compiled and tested independently. SMALLTALK-80 is an example of a class of so-called 'object-oriented' languages which are designed to facilitate problem description by, for example, allowing a newly defined object to 'inherit' some of the characteristics of previously defined objects.

System components. The main conceptual development involved is the idea that, like data, the rules are separated from the rest of the program. The collection of rules is known as the knowledge base; these fit into what is known as a shell. This separation of the knowledge (which may be represented in a variety of ways) from the mechanism for manipulating the knowledge makes it very much easier to introduce new rules, and to see the structural relationships involved in the knowledge base, compared say with a Fortran program to do the same job. The other main feature of an expert system is the mechanism involved in the manipulation of knowledge and drawing of inferences. Technically, this part of the expert system is known as the inference engine; some of its characteristics are described below.

Uncertainty in input. Expert systems can deal with imprecision in input data either by requiring the user to specify probability weights for different states of the input variables or, more rarely, by the use of sophisticated procedures based on fuzzy logic.

Uncertainty of outcome. While some inference mechanisms are deterministic involving what are know as 'production rules' (of the form 'IF A is true THEN B is true'), most mechanisms involve statistical procedures in order to make inferences. Thus for example the rule might be 'IF A is true THEN B will be true with probability 0.7 AND C will be true with probability 0.3'. Although there are some alternative techniques available, each with their own advocates (see Mandani et al, 1985), most expert systems use the well-known statistical principle of Bayes' Theorem to manipulate the probabilities in the rule set so as to reach an overall judgement of the probability of a particular objective being true. This consistency and the appropriate weighting of probabilities is an important attribute of such systems.

Explanation and description. An important attribute of an expert system is that it can be asked to explain its reasoning. Since it is desirable, for ease of comprehension, that its rules and explanations are readily recognised, the computer has an enhanced descriptive role. The programmer will therefore have to spend a greater proportion of his/her effort on ensuring that the descriptive text and rules are readily understood, and that the system is developed in a user-friendly way.

THE SCOPE FOR APPLYING EXPERT SYSTEMS IN TRANSPORT

Context.

A number of authors (eg Wigan, 1983, Logie and Neffendorf 1984) have drawn attention to the possibility of applying expert systems to aid the study of transport problems. The work reported here is the outcome of a detailed examination of the scope for such applications, carried out at the Institute for Transport Studies (ITS) of the University of Leeds. The range of applications considered, though wide, naturally reflects the particular interests of the Institute; others considering a breadth of transport applications include Wigan (1985) at the Australian Road Research Board, Hendrickson et al (1985) at Carnegie-Mellon University, Philadelphia, and Yeh et al (1985) at the University of Washington, Seattle. Many other institutions are now responding to the challenge of adopting AI techniques for use in transport applications, particularly in the USA; this is indicated by the way in which the topic bubbles up in the papers reviewing new transportation research opportunities (Boyce, 1985). References to other applications are included below.

The following list summarises potential applications of expert systems considered in the ITS project. It should be noted that the list is arranged by type of application and therefore cuts across the technical distinctions between different types of expert system.

Enquiry Systems

- advice on regulations
- travel enquiries
- route advice
- advice on data sources

Analytical Advice and Interpretation Systems

- procedural and methodological advice
- interpretation

Design Systems

- infrastructure design
- network/junction design

- schedule design
- questionnaire design

Diagnostic/Prescriptive Systems

- road safety systems
- road maintenance
- structures and equipment maintenance

Identification Systems

- inventory and condition logging
- automatic identification

Control Systems

- traffic monitoring and control
- air traffic control
- survey control

Policy Support Systems

- multi-criteria decision making
- treatment of uncertainty
- consistency of policy

We now consider what might be involved in each of these applications. Because of the range of subjects covered, the different applications are discussed in different levels of detail, but it is hoped this will stimulate further debate.

Enquiry Systems.

The 'expertise' inherent in a successful enquiry system is the ability to respond to an enquiry so as to provide the necessary information as quickly as possible. This implies an ability to tease out of the enquirer precisely what it is that he wishes to know, to access the relevant information and to pass it on clearly and succinctly. Of these the most difficult is the interpretation of the initial enquiry and the decision as to what information would be useful. In order to provide such a service it is of course necessary to have a sound and up-to-date knowledge of the field in question.

Expert systems have much to contribute here because of their ability to arrange their knowledge bases such that they 'know' the extent of their knowledge (a concept known as meta-knowledge), because their knowledge can be rapidly updated from a variety of sources, because they can respond to questions posed in a variety of ways, because they can respond flexibly (eg changing direction in response to additional or amended information from the enquirer) and because they can learn from experience (what is found useful in one session may be useful in similar circumstances in a subsequent session).

Another feature of expert systems that might commend their use in enquiry systems is their ability to act through a natural language interface whether typed or even spoken. Travel enquiries have often featured in trials of dialogue interpretation programs eg GUS (described by Bobrow et al, 1977) but in practice the complexities inherent in such systems may rule them out and leave menu selection as the most efficient mode of dialogue (see Rich, 1983 p 334).

Enquiry systems can of course be designed either as a complete service for lay enquirers, as training aids for novice clerks or as reference aids for more-or-less experienced clerks. In the latter case it is clearly important that the expert system is designed to provide accurate information about complex or infrequently requested topics more quickly than would be possible via more conventional means. Experience with enquiry systems based on conventional computing languages and data-base management systems has shown that, despite the benefits of having the necessary information organised in a database, the updating and accessing of the system may be so frustratingly tedious as to lead to such systems being ignored. There are, of course, several types of enquiry system - we mention only four. They range from simple advice systems that can be developed in a self-contained shell, to those that inter-relate with databases.

Advice on regulations. An expert system whose knowledge base contained regulations (eg building or planning regulations, vehicle construction-and-use regulations, customs regulations) would be relatively easy to set up since the knowledge is already codified. In its basic form, the main advantages of such a system over a physical book of rules are the ease with which it can be updated to include new or amended regulations, and the explanations it can provide on its recommendations. A more ambitious system might incorporate case study and precedence information and might build up its own case law (seeking the important circumstances and antecedent regulations from its informants). Such a system might then be able to quote precedence as well as regulatory justification for its advice in given circumstances.

Travel enquiries. The intelligent front-end to a travel enquiry system ought to include the following features:

- optimum organisation of information so as to speed up the enquiry (eg with the system learning, from its experience with previous enquiries, that before holiday periods a high proportion of enquirers are likely to be interested in discount fares);
- ability to deal with requests posed in a variety of ways (eg 'I want to leave after x', 'I want to arrive by y', 'I want to travel in daylight' ...);
- ability to deal with fuzzy requests (eg 'what trains travel before that?');
- ability to devise sensible trade-offs (eg 'that train will get you in at 11 am', 'if you set off 5 minutes earlier you could get in at 10am');
- ability to explain in an alternative format any instructions not immediately understood (eg to give times in terms of the 12-hour rather than 24-hour clock);
- ability to provide subsidiary information if requested (eg whether changes of platform are likely to be required).

An additional benefit of expert systems in travel enquiries would be an ability to combine data from a variety of sources (eg timetables in a variety of formats from a variety of operators, temporary timetable changes, special discount fares, real time information on delayed services etc).

Our own experience at the Institute for Transport Studies in designing a simple railway timetable and fares enquiry system (TALK: Train Advice Leeds-Kings Cross) using a SAGE shell on a UNIX VAX computer convinced us of the limitations of the shell's facilities (see Kwan, 1985) and led us to propose a more ambitious system to be developed via the Alvey Directorate's IKBS Awareness Programme (Software Sciences et al, 1985). As a result, the TRACE Community Club has been formed to produce prototype enquiry systems not only for public transport services but also for holiday and travel sales. (1)

Route advice. A number of conventionally constructed route advice systems already exist, some for private vehicles (eg those provided by automobile associations for their members) and some for public transport users (eg Williamson and Miller, 1981). These are based on minimum 'cost' path-finding algorithms and thus an amendment (eg to take account of temporary changes or interruptions to the network) is relatively expensive to incorporate. An expert system adviser might overcome this problem by amending only those routes which it knows to be affected by the change. Other desirable features might include an ability to define routes according to a variety of user-specified criteria (eg time, distance, reliability, avoidance of low bridges, etc), to define routes in the user's preferred terminology (eg 'take signs for Oxford' or '3rd left, then 2nd right...'), to provide subsidiary information if requested (eg places of interest en route), to provide trade-off information (eg 'the route that best meets your instructions is ... however, if you are prepared to go through Basingstoke you could save 15 minutes') and to learn from experience (eg that people do not want to go through Basingstoke no matter how much time it saves!). Most of the features can, of course, be provided via conventional software (see for example Wootton and Brett's (1986) description of the UK Automobile Association's new system) and it remains to be seen whether the potential benefits of an expert system version (notably easier updating and learning from experience) are worthwhile. (2)

Routing advice for multi-drop tours is discussed below under 'schedule design'.

Advice on data sources. Users of transport statistics have focussed attention on the widespread difficulty they have in becoming aware of the existence, relevance and availability of appropriate secondary data for a specific purpose (McLean, 1983). Although keyword indexing and computerised data bases have been of significant importance they clearly leave something to be desired in their habit of unearthing either 'too much' or 'not enough'. An expert system might be able to mimic the ability of a knowledgable, efficient (and patient!) statistical sources expert or librarian to tease out of the enquirers precisely what they wish to know and to point them to the relevant data. A proposal to prototype such a system is currently under development at the Institute for Transport Studies.

Analytical Advice and Interpretation Systems.

Advice on technical procedures/methodologies, covering such matters as choice of appropriate survey techniques, statistical tests or types of model, would obviously rely on precedence and established procedures. At their simplest, such systems might appear little more than a computerised checklist of techniques or methodologies that have been regarded as suitable in similar circumstances. However, the main advantages of a more advanced expert system would be that it might probe for the determining circumstances and, in learning mode, might learn from the user's reactions to its proffered advice. Also, the system could, on the basis of accumulated experience and practice, take account of tradeoffs between any extra accuracy of more ambitious procedures and the extra cost and data requirements which might be incurred. Clearly, the advice system could go further than simply suggesting which techniques should be used; it could also give advice, if asked, on how they should be used. For example, a session with a sampling strategy advice system might begin with the user making statements on survey budget, desired confidence levels and anything known on the shape of the distributions to be sampled. The system might then ask for additional information (eg on network configuration) before suggesting the numbers of counts or interviews of various types to be carried at specified locations at specified times. The user might then respond on the feasibility of the suggested strategy (from which responses the system could learn for future reference).

Interpretation systems. An expert system might be designed not only to provide advice but, through its interface with the analytical procedures in question, might actually implement the advised procedures. Thus a system advising on the choice of statistical tests or analytical models might be designed to analyse the results of such models and hence to suggest a series of tests or models each based on the results of the previous one. (Thus, in the transport demand field, a statistical test on mode choice data might suggest that a lexicographic mode choice model be run in preference to a compensatory model). Particular attractions of such a system are that it would ensure consistency between models and the parameters/ coefficients which they share and would clearly save the analyst considerable hours of work poring over voluminous computer output.

It may seem a logical progression to move from an advice system such as those outlined above to a system which might, without pausing for human intervention, requisition the data, carry out the tests and implement the models which it thinks appropriate. Whether this is a desirable goal must depend on whether the 'rules' which govern such actions can be specified. A halfway house would be an interpretation system whose task was just to assist the human in his assessment of information in data-bases and in model outputs (drawing his attention to the most significant features) and which, if designed to accept data in a variety of formats, could simplify the provision of information required by otherwise free standing analytical procedures.

Design Systems.

The design process is a complex activity, meaning different things in different contexts. The outcome is eventually an arrangement of objects to perform a given task and meeting certain criteria. Computer-aided design has hitherto embraced just the drafting aspects of this, concentrating on the use of interactive computer graphics. There is considerable scope for extending the computer's role via AI techniques to cover the selection and examination of alternative designs that meet the appropriate criteria (see Mostow, 1985).

Three features of expert systems that might prove particularly valuable in analytical design systems are:

- an intelligent front end to established design procedures that could deal with information presented or required in a variety of formats;
- a readily modified rule base that could limit the range of designs to be evaluated (eg through specific safety or capacity requirements or through constraints on project budget or duration); and
- an intelligent interpreter of the output of design exercises that could summarise and evaluate the important strengths and weaknesses of particular designs.

We now consider how four different design systems might benefit from the application of expert systems technology.

Infrastructure design systems. The knowledge base for designing infrastructure would include information on the components, how they relate to each other, what overall criteria should be met and what measures of performance should be calculated. Examples include bridge design (for a pilot expert system on which see Welch and Biswas, 1986), car-park layout, and loading-bay configurations. System features might include:

- user specification of major parameters such as cost, capacity, maximum overall dimensions, site configuration;
- user specification of special features and constraints (eg that it must be constructable within 6 weeks, that there is a supply of cheap aggregate available nearby);
- draft designs suggested by the system (in learning mode constructive user comment on the draft designs would ensure enhancement of their standard in future applications);
- user selection of preferred design;
- system requests for further information as necessary;

- production of final design.

Engineering calculations (eg of stress for a bridge, or of curvature for a road) would need to be embodied within the overall design process as would user-friendly interactive graphics features.

Network/junction design systems. Such systems would need to consider not only the dimensions of space and time but would also have to reflect the fact that facility usage may be dependent on the design (eg traffic flows being affected by road network design). Interfacing with traffic/travel demand models and data will therefore be an important feature of the overall system.

An example of the use of an expert system as a design aid is the experimental program TRALI, developed at Carnegie-Mellon University. TRALI serves as a 'traffic signalling assistant' by suggesting sequences of phases for an isolated traffic signal (Zozaya-Gorostiza and Hendrickson, 1985); its expert system features are mainly in respect of rule-based limits on the range of permitted designs. Further work at Leeds is expected to develop this approach; the main advantage is that flexibility is facilitated by having the rules in a readily modified rule-base rather than subsumed in the code of more conventional traffic signal optimising and sequence design programs.

Schedule design. Routeing for a multi-drop delivery is of course the 'travelling salesman' class of problem. For practitioners the problem is compounded with that of scheduling, because different vehicles in the fleet have different availabilities and capacities. In practice, schedulers are likely to rely on empirical rules of thumb rather than on algorithms, as these have hitherto been rather clumsy and time-consuming to use - especially if constraints are imposed on the time at which drops may be made. A knowledge-based approach, probably coupled with algorithms, may provide a useful way of harnessing an experienced scheduler's skills and making them more widely available in an organisation, if implemented on a micro-computer. Golden and Baker (in Boyce, 1985, p 407) and Polak and Jessop (1986) have also drawn attention to the potential role of expert systems in logistics.

The scheduling of buses and bus crews raises similar issues. In both the public transport and the freight situation, the knowledge base would need to incorporate experience from previous practice, and knowledge of local working practices and union agreements.

3) Questionnaire design systems. We have considered whether there might be a role for expert systems in questionnaire design and conclude that, to be useful, they would require the following features:

- user specification of the items of data sought;
- system suggestion of draft questionnaire wording (drawn from a menu of possible questions stored in its knowledge base): the system would previously have checked which items are to be asked of which respondents in which circumstances;
- in learning mode, the system might invite user comments on the format of the proposed questions and their relationship one to another: any comments being incorporated into the system's knowledge base. The system might, if it came across a data item it had not yet met, or if its advice on wording were criticised, ask the user to provide a suggested wording;
- a substantial part of the system would be concerned with ensuring that the component questions fit properly together: user feedback on this would again allow the system to build on its own experience.

However, unless one were designing a system with highly specific objectives (eg how to ask a question about income), it seems most unlikely that a system could satisfactorily cope with the typically very varied nature of questionnaires and subject matter. A system based in conventional computing, such as QUESTMAST (Centre for Educational Sociology, 1984), may well be a sufficiently helpful design aid for survey documentation, coding manuals, variable definitions and specification of statistical analyses.

Diagnostic/Prescriptive Systems.

The scope for simple expert systems to advise on issues such as road maintenance treatment and environmental assessment has been indicated by several authors (see for example, Wigan 1982 and 1985; Hendrickson et al, 1985; Harris et al, 1985). Appropriate features would be:

- an ability to assimilate data from a variety of sources, to diagnose problems, preferably in advance of actual failure, and to suggest cost effective solutions to avoid any (further) failure;
- calculation of the probabilities of failure/accident if given remedial measures are (or are not) taken;
- a learning mode wherein human experts feed in rules of thumb, case histories and other information;
- a knowledge updating mode wherein the system monitors the performance of its own prescribed solutions.

Examples include:

Road safety systems. The aim here would be to diagnose potential accident blackspots and hazards and prescribe remedial measures. Data input might include accident case histories and site characteristics (infrastructure, traffic, weather). The site characteristics data might well be continually updated from on-line databases. An investigation of such systems is currently underway at University College London.

Road maintenance systems Data inputs for these might include structural history (age, materials used, strata thickness, ground conditions, previous maintenance) cumulative traffic flow (by weight), weather conditions and (in a sample of locations) embedded instrumentation to provide continuous data on porosity, compaction, deformation etc.

The integration of instrumentation with on-line monitoring through the medium of expert systems has been suggested by Moavenzadeh (in Boyce, 1985, p 505). and pilot work on a small-scale adviser is being conducted in the Department of Civil Engineering at the University of Leeds.

Structures and equipment maintenance systems. Data inputs for these would be analogous to those described for road maintenance systems. Clearly systems with this remit would normally be produced from within the structural and mechanical engineering fraternity. Nevertheless, there are several examples of structures and equipment which are particularly oriented to transport (eg bridges, rail track, vehicles, etc) and which might warrant specialised systems. Wood (1985), for example, has discussed an expert system for bus maintenance; the US Army have developed an expert system for advising on railway track maintenance.

Identification Systems.

Inventory and condition logging systems might provide an aid to relatively unskilled enumerators engaged in inventory and condition surveys (or even highly skilled enumerators engaged in highly complex inventory and condition surveys). They would enable the enumerator to describe the inventory item in his own terms and the system would then prompt him with appropriate questions to determine exactly what was being described and what its condition was.

The essential features of the system would be:

- acceptance of a variety of imprecise definitions in free text;
- determination of the most significant questions so as to home in on the precise nature of the item being described (be it height, colour, serial number etc);
- ability to be used on site or in the field;
- a learning mode within which the system could learn new terms (eg lamp standard = lamp post) and learn to accept imprecision where precision is unnecessary (a 'yellow' light might not need to be distinguished from an 'orange' one for some purposes);
- easy updating of the knowledge base to include items newly introduced (eg a new type of road sign) or not yet met (eg an obsolete type of traffic light), or to allow for new knowledge (eg that concrete spalling is preceded by such-and-such a symptom).

The advantages of such a system over a decision-tree based algorithm would be its flexibility and efficiency, but again, it is not clear whether these advantages would justify the extra effort required.

Automatic identification systems such as image processing, voice recognition or recognition of patterns from inductive loops are currently being developed to aid the identification of classes of vehicle, items of inventory and even individual vehicles. They could each be enhanced by being linked with expert systems. The linkage between pattern recognition and IKBS is a contentious issue of fundamental research well beyond the scope of this paper, but even without recourse to such techniques it is clear that an expert system could help resolve ambiguities in the output of an automatic identification system. In essence such aid would amount to the system saying "in the circumstances it is probably x". The circumstances in question could cover whether:

- anything is known about the population being observed (eg at a previous observation point a given mix of vehicles might have been observed);
- anything expected/unexpected in the population (eg one might not expect to see a fire engine in a multi-storey car park or a fire hydrant on a motorway).

The training of such systems could obviously start by incorporation of rules of thumb (eg local registration plates predominate during peak hours) and would be an on-going process whereby the data base of contextual information linked with successful identifications is continuously updated. (4)

Control Systems.

This group of applications are characterised by their use of on-line data to optimise the performance of a system. They would derive their knowledge of how to achieve that optimisation from encoded formulae and relationships supplemented by case histories including those to which they have input advice.

Traffic monitoring and control is an obvious example of such an application. Most of the 'advice' would doubtless best be effected automatically (eg rephasing of signals in a UTC system) but it might be dèsirable sometimes to issue advice that could only be effected by human intervention (eg "check a particular stretch of road for physical obstruction"). The data input to the expert system would be most obviously via automatic sensors (eg flow and queue detectors) but one can imagine an expert system which was able to request extra items of data from roving human enumerators in order to assist in the resolution of a given problem (the system might, for example, request that the length of the queue at a certain uninstrumented junction be measured or that a sample count be taken at a location where the automatic counter is defective). The decision to request such information, just like any decisions made by the expert system, would be based on its desire to optimise the performance of the traffic system subject to the costs of so doing being lower than the probable benefits to be derived. This again is an area where an expert system that could learn from its own experience and the constructive comments of human experts would be
5) particularly helpful.

Air traffic control has so far attracted more attention as a possible subject for using AI techniques than has road traffic control. In the USA, proposals to develop advanced automation systems for air traffic control include new capabilities (known as AERA) for en-route detection and resolution of control problems; these will provide the controller with additional planning information and the optional automative features, both to help in the near-term (within 20 minutes) and longer-term time horizon (Elsaesser et al, 1984).

Though the AERA proposals do not explicitly envisage the use of AI techniques, their potential is being investigated through some experimental systems. In presentations at the 1986 Meeting of the USA Transportation Research Board, possible applications of AI helpful to air traffic control were identified as: detection of collision-risk situations; suggestion of collision-avoidance manouevres; prevention of aircraft build-up; provision of advance information on aircraft movements; management of displays; advice on flight service procedures; checks on equipment safety; assistance in training; and detection of freak weather conditions.

A pilot example of the first of these, collision avoidance, was developed by the MITRE Corporation. Known as AIRPAC (Adviser for Intelligent Resolution of Predicted Air Conflicts), this considered conflicts between just two aircraft, and involved four succesive stages: problem decomposition; tactic selection (the kernel of the system); tactic development; and manoeuvre parameter selection. The Lincoln Laboratory have developed a system for detecting wind-shear patterns from weather radar. In the UK, the Civil Aviation Authority is considering the application of IKBS techniques to flow regulation (to minimise delay), and to the use of shared air-space.

Notwithstanding these developments, it seems unlikely that expert system technology will completely replace the human air traffic controller, if only because of the suspicion that no fully automatic system will be as able as a human to cope with the unpredictable. Expert system technology is, however, likely in due course to provide valuable decision aids. (6)

A survey control system might be linked to on-line data collection devices or might have data transmitted to it periodically from data loggers in the field. With this information it would be able to calculate distributions and variances and, knowing the desired level of precision for given data items, would then issue advice as to what extra data ought to be collected where and when. Use of the system would then ensure minimum waste of survey effort. Much of this could, of course, be achieved by conventional computing but an expert system might be better able to act on contextual information such as overall project budget, site conditions, possible trade-offs between the desired precision of different data items, marginal costs of extra data collection etc. The advice system might learn how to deal with this contextual information by receiving constructive comments from human experts on its suggested actions and by monitoring the success or otherwise of its own suggestions.

Policy Support Systems.

Expert systems can in general be regarded as decision aids, and there is a rapidly developing literature on their application in the context of decision support systems. We include here just three examples of how they might assist in a policy support role.

Multi criteria decision making involves assessing the relative value of options which are described in terms of multiple criteria - eg two alternative routings for a new road might give quite different benefits in terms of time savings, accidents, environmental effects, etc. An aid to such decision making might be provided by an expert system which, on the basis of previously logged decisions or of trade-off exercises set by the system, could devise the most appropriate way of combining the different criteria. Trade-off exercises might involve the user being asked to rank or score various packages of attributes or, more directly, to specify trade-off values (eg a 1% reduction in road accidents is of equivalent value to a 2% increase in journey time). Such a system would mimic the procedures currently recommended by professional decision analysts.

Advice on the treatment of uncertainty in forecasting or appraisal might be particularly useful. The system might calculate the net costs and benefits of a series of possible outcomes and then, having ascertained the policy maker's attitude to risk on the basis of direct questions or by deduction from his past decisions, would advise on the optimal decision. Clearly much of the benefit of such a system might be obtained from conventional computing (why has it not been done?) and it is perhaps only the ability to deduce the decision maker's attitude to risk from his past decision that is characteristic of an expert system (although even this might be achieved via more conventional techniques).

Advice on the consistency of policy might not always be welcomed by a policy maker but he might accept it more readily from a machine than from a human! A system might well be developed to investigate past policy decisions and deduce values put on various items (eg the value of commuter travel time, of human life, of clean air, of aesthetics etc). Once this has been done the system might either suggest what decision on an outstanding question would be consistent with earlier decisions or might point out inconsistencies between the proposed decision and earlier ones. The system might seek the reasons for such apparent inconsistencies and hence update its knowledge base. Thus it might say 'the proposed decision implies a much higher value on human life than any previous decision, do you mean this to be the case and if so how do you justify it?'.

DISCUSSION

The successful application of AI and expert systems techniques in transport, and indeed in any applied discipline, will depend on the ease with which transport professionals adopt and use the techniques for themselves. Unfortunately, there are at present several barriers to understanding such systems. The jargon is most off-putting, there is a lot of it, several terms may be used for the same thing, and the same term may mean different things to different people. Concepts are different, and the way of thinking about a problem is different. Illustrative material often seems trivial or irrelevant.

Fortunately, a number of texts are becoming available to help de-mystify and explain the subject; guidance on some of these is given in the Appendix.

In course of time the techniques will probably become as much a part of the armoury of tools as statistical methods are now. However, unlike statistical methods, familiarity with the tools cannot be acquired in isolation from the computer. The ways in which the computer represents knowledge and relationships are intrinsic to the tools, and this means that their successful adoption will require much dedicated effort.

The basic concept of embodying human rules of thumb and expert judgement in a computer program is straightforward to grasp but difficult to apply. It is all too easy to fall into the trap of expecting too much of an expert system. Expert systems have outperformed humans only for very strictly defined tasks. More often they are able only to mimic the human's performance in a quite restricted part of his domain of expertise.

Many of the early expert system shells came about simply by stripping out the original knowledge base from a given expert system. For example, the shell EMYCIN came from MYCIN (here "E"="Empty"!). It is important to realise that a shell designed for one purpose may be quite inadequate for another. In particular, some of the shells treat uncertainty in a quite simplistic manner.

In some fields in civil engineering, such as construction and building, regulations provide rich scope for straightforward applications of expert systems using relatively simple shells on micros. They thus provide a good industry-wide starting point for familiarisation with what such systems can do. Similarly simple transport applications, on the other hand, are rather restrictive in their appeal and the transport applications we have identified above are likely to be rather demanding of expert system technology. In this respect we do not accept d'Agapeyeff's (1984) conclusion that "simpler expert systems are practical" for most applications of interest in the transport sector.

Serious transport applications will place several demands on the developers of expert system technology. Foremost, given the importance of number crunching in most transport applications, must be the development of systems or languages that are much better able to integrate the handling of numbers with rules and relationships. Secondly, since systems fall into disuse if cruder but quicker techniques will suffice, expert system technology needs to be speeded-up; not only for real-time applications such as traffic monitoring and control, but also more generally. But perhaps the greatest challenge facing the developers of expert system technology is to improve the systems' ability to learn from their experience in being used. Several times in our list of transport applications we have suggested that this ability is particularly beneficial; but it is a strange anomaly of the artificial intelligence field that, though simple applications of such 'automated learning' processes can be found in most introductory AI texts and even programmed in Basic for home micros, the introduction of automated learning techniques in serious applications is extremely difficult, and generally not provided for in current software tools.

Alongside the need to develop expert system technology itself, is the need to develop the man-machine interface via enhanced dialogue facilities. It is similarly necessary to develop the skills and procedures for acquiring the knowledge that is to be built into the system. This need lies at the heart of any expert system, but surprisingly little progress has been made in the development of systematic knowledge acquisition methodologies (for reviews on which see Welbank, 1983, and Bourne and Sztipanovits, 1985). One implication of this is that the development of an expert system should not be left solely to an expert system specialist; general knowledge of the subject will be important in identifying the right sort of questions to ask, of whom and in what order.

Expert system technologies provide the tools to enable a wider range of problems to be tackled (or to enable a given problem to be tackled more comprehensively). The goal must not be to produce "expert systems" per se; but rather to produce systems in which expert system tools are used for what they are good at, and algorithmic procedures used for what they are good at. Arguments as to whether one should use 'shells' or 'languages', or whether the eventual program is worthy to be called an expert system or not, will eventually be seen to be rather sterile. Expert system technology will in the end only be thought useful if embedded in an approach that confronts the whole of a problem, not just part of it.

Whatever the end product be called, the need remains for practitioners and clients to become better acquainted with what expert system methodology can do, and for their analysts to become well versed not only in what it can do, but how different tools contribute to doing it. The development of pilot, but not trivial, demonstration projects should contribute significantly to this end.

APPENDIX: SOME INTRODUCTORY SOURCES

The need to de-mystify and explain the subject of artificial intelligence and expert systems is now being addressed at several levels. Home computer users have become greatly interested in the techniques, partly perhaps because some of the AI techniques (such as text compression and search procedures) are exploited in the advanced adventure games now available. This has resulted in a variety of easy-to-use introductory sources on the subject, covering not only books (eg Naylor, 1983) but also viewdata-based user groups (MICRONET, 1986) and useful, though somewhat limited, implementations of LISP and micro-PROLOG on home micros.

At a more serious level, there are a number of introductory articles in journals for specific application areas; see for example our own report (Wheatley, 1984) for transport, and Hinde (1985) for Operational Research. Several good text books are available describing the essential background to and nature of the subject: see for example Rich (1983) and Bonnet (1986). For those wanting to get the feel for commercial expert systems, some 'starter packs' are now available, ranging from those that cope only with production rules to those that incorporate scaled-down versions of commercial systems; that issued for the Alvey Directorate (1984) by the National Computing Centre contains what Wigan (1985) describes as 'crippled' versions of Expert-Ease, ESP Adviser and Micro-Expert, and Micro Synics. Useful reviews of many of the leading expert system shells can be found in Allwood et al (1985) and Wigan (1985); summary details can also be obtained from the Institute on request. A useful catalogue of a wider range of artificial intelligence tools is given by Bundy (1986).

REFERENCES

d'AGAPEYEFF, A. (1984) Report to the Alvey Directorate on a short survey of expert systems in UK business. Alvey News, Supplement to Issue No. 4.

ALLWOOD, R.J., STEWART, D.J., HINDE, C. and NEGUS, B. (1985) Evaluation of expert system shells for construction industry applications.
Department of Civil Engineering, University of Lough-borough.

ALVEY DIRECTORATE (1984) IKBS Starter Pack. National Computing Centre, Manchester.

BOBROW, D.G., KAPLAN, R.M., KAY, M., NORMAN, D.A., THOMPSON, H. and WINOGRAD, T. (1977) GUS, a frame-driven dialog system. Artificial Intelligence, 8, 157.

BONNET, A. (1986) Artificial intelligence. Promise and performance. Prentice Hall International, London. ISBN 0-13-048869-0, £9.95.

BOURNE, J.R. and SZTIPANOVITS, J. (1985) Strategies for knowledge representation, manipulation and acquisition in expert systems. Seventh Ann. Conf. of the Engineering in Medicine and Biology Society, Institute of Electrical and Electronic Engineering, 1165-1169.

BOYCE, D.E. (ed) (1985) Transportation research: the state of the art and research opportunities. Transportation Research, 19A(5/6), Special Issue.

BUNDY, (1986) Catalogue of artificial intelligence tools. Springer. ISBN 3-54-013938-9, £13.25.

CENTRE FOR EDUCATIONAL SOCIOLOGY (1984) QUESTMAST users' manual. Centre for Educational Sociology, University of Edinburgh.

DEPARTMENT OF INDUSTRY (1982) A programme for advanced information technology. The Report of the Alvey Committee. HMSO, London, ISBN 0-11-513653-3, £4.30.

ELSAESSER, C., GISCH, A.H., HAINES, A.L. and SWEDISH, W.J. (1984) Description of AERA 1 capabilities. MTR-84W88, Metrek Division, Mitre Corporation, McLean, Virginia, USA (unpublished).

FEIGENBAUM, E.A. and McCORDUCK, P. (1984) The fifth generation. Artificial intelligence and Japan's computer challenge to the world. Pan Books, London.

HARRIS, R.A., COHN, L.F. and BOWLBY, W. (1985) An application of artificial intelligence in highway noise analysis. Transportation Research Record 1033.

HENDRICKSON, C.T., REHAK,D.R. and FENVES, S.J. (1985) Expert systems in transportation systems engineering. Research Report, Department of Civil Engineering, Carnegie-Mellon University.

HINDE, C.J. (1985) Artificial intelligence and expert systems. In: Rand, G.K. and Eglese, R.W. (eds) Further developments in Operational Research. Pergamon Press, Oxford.

KWAN, R.S.K. (1985) Approaches to expert systems for passenger enquiries using SAGE. Techical Note 163, Institute for Transport Studies, University of Leeds, Leeds.

LOGIE, M. and NEFFENDORF, H. (1984) Expert systems in transportation. PTRC Summer Annual Meeting, Brighton 1984.

MACLEAN, I. (1983) Proceedings of the Transport Statistics Conference, November 1983. Annual Conference of the Statistics Users Council. IMAC Research, London.

MANDANI, A., EFSTATHIOU, J. and PANG, D. (1985) Inference under uncertainty. Unpublished report, Queen Mary College, University of London.

MICRONET (1986) AI Queries. PRESTEL, frame *8006055 et seq.

MOSTOW, J. (1985) Towards better models of the design process. The AI Magazine, Spring 1985, 44-56.

NAYLOR, C. (1983) Build your own expert system. Sigma, Wilmslow, ISBN 0-905104-41-2, £6.95.

O'KEEFE, R.M., BELTON, V. and BALL, T. (1985) Getting into expert systems. QSS Discussion Paper No. 71, Board of Studies in Quantitative Social Science and Management Science, University of Kent at Canterbury.

POLAK, J.W. and JESSOP, A. (1986) Towards expert systems for goods distribution management. Conference Universities Transport Study Group, held at UWIST, Cardiff, Jan. 1986 (unpublished).

RICH, E. (1983) Artificial intelligence. McGraw-Hill.

SOFTWARE SCIENCES Ltd, WOOTTON JEFFREYS plc, and INSTITUTE FOR TRANSPORT STUDIES, UNIVERSITY OF LEEDS (1985) Expert enquiry systems. A proposal to the Alvey Directorate to form a Community Club in Transport. Software Sciences Ltd, Farnborough.

WELBANK, M. (1983) A review of knowledge acquisition techniques for expert systems. British Telecom Research Laboratory, Martlesham Consultancy Services, Ipswich.

WELCH, J.G. and BISWAS, M. (1986) Application of expert systems in the design of bridges. 65th Annual Meeting, Transportation Research Board, Washington DC, January 1986.

WHEATLEY, M.D. (1984) Expert systems in transport. Part 1: An introduction to expert systems. Working Paper 178, Institute for Transport Studies, University of Leeds, Leeds.

WIGAN, M.R. (1983) Information technology and transport: what research needs to be started now? Working Paper 172, Institute for Transport Studies, University of Leeds, Leeds.

WIGAN, M.R. (1985) Knowledge based systems tools on microsystems. Research Report ARR 134, Australian Road Research Board, Vermount, Victoria, Australia (43pp). ISBN 0-86910-206-0 (Report); ISBN 0-86910-207-9 (Microfiche).

WILLIAMSON, R.H. and MILLER, A.J. (1981) A personalised public transport directory. Annual Public Transport Symposium, University of Newcastle upon Tyne, July 1981.

WOOD, P. (1985) A knowledge based system for transit bus maintenance. Transportation Research Record 1019, 85-91.

WOOTTON, H.J. and BRETT, A.C. (1986) Route information systems - signposts of the future? Proceedings of the Second International Conference on Road Traffic Control, Institution of Electrical Engineers, April 1986, London.

YEH, C-I., RITCHIE, S.G., SCHNEIDER, J.B. and NIHAN, N.L. (1985) Knowledge-based expert systems in transportation planning and engineering: an initial assessment. Annual Meeting, Transportation Research Board, Washington D.C., January 1985.

ZOZAYA-GOROSTIZA, C. and HENDRICKSON, C. (1986) An expert system for traffic signal assistance. Research Report, Carnegie-Mellon University.

EDITORS' NOTES

(1) For a full discussion of more conventional methods of responding to public transport enquiries see Chapter 13, and for a review of the role of communications systems in the travel industry, see Chapter 12.

(2) A comprehensive review of route guidance systems is provided in Chapter 14.

(3) Chapters 9, 10 and 11 deal in some depth with scheduling techniques currently in use in the freight and bus industries and with the problems which have to be surmounted.

(4) See Chapters 2 and 4 for an insight into existing identification and recognition systems.

(5) Chapter 6 describes a number of the systems, based on conventional computer languages, which address the question of traffic monitoring and control.

(6) Chapter 8, which is concerned with air traffic control systems, suggests that there has been some reluctance among ATC personnel to accept more conventional computing aids, let alone expert systems.

CONTRIBUTING AUTHORS

Names	Addresses
ASHWORTH Robert	Dept of Civil and Structural Engineering, University of Sheffield, Mappin Street Sheffield, S1 3JD, UK.
BELL Michael	Transport Operations Research Group, University of Newcastle upon Tyne, Newcastle upon Tyne, NE1 7RU, UK.
BONSALL Peter	Institute for Transport Studies, University of Leeds, Leeds, LS2 9JT, UK.
CATLING Ian	Transpotech Ltd, 4th floor. Thames House North, Millbank, London SW1P 4QE, UK.
CORMIER Michel	Centre de Reserche sur les Transports. Université de Montréal CP 6128 Succursale A, Montreal. Quebec H3C 3J7, Canada.
DAVIES Peter	Department of Civil Engineering. University of Nottingham, Nottingham NG7 2RD, UK.
DICKINSON Keith	Department of Civil Engineering, Napier College, Colinton Road, Edinburgh, EH10 5DT, UK.
GEEHAN Tom	TransVision Consultants Limited, 3230 Chaucer Avenue, North Vancouver BC, V7K 2C3, Canada.
HOLT John	Operations Development Department, British Rail Board Headquarters, 222 Marylebone Road, London, NW1 6JJ, UK.
JEFFERY David	Transport and Road Research Laboratory, Crowthorne, Berks, RG11 6AU, UK.
KIRBY Howard	Institute for Transport Studies, University of Leeds, Leeds, LS2 9JT, UK.
KLIJNHOUT Job	Rijkswaterstaat, Koningskade 4, Postbus 20906, 2500 EX's-gravenhage, Netherlands.
LAPALME Guy	Department d'Informatique et de Recherche Operationelle, Université de Montréal, CP 6128, Suggursale A. Montreal, Quebec, H3C 3J7, Canada.
MELLOR Andrew	Halcrow Fox Associates. Vineyard House, 44 Brook Green, Hammersmith, London W6 7BY, UK.

POPE Richard — Fir Tree Consultancy Ltd, 4 The Coopers, Itchingfield, Horsham, West Sussex, RH13 7PQ, UK.

ROSS Nigel — Software Sciences Limited, Farnborough, Hampshire GU14 7NB, UK.

SAHLING Michael — PTV GmbH, Rintheimer Strasse 48, D-7500 Karlshruhe 1. Federal Republic of Germany.

SUEN Ling — Transportation Development Centre, Transport Canada, suite 601,West Tower, 200 Dorchester Blvd. West, Montreal, Quebec. H2Z1X4, Canada.

WATERFALL Roger — Department of Electrical Engineering and Electronics, UMIST, Sackville Street, Manchester. M60 1QD, UK.

WREN Anthony — Operational Research Unit, University of Leeds, Leeds, LS2 9JT, UK.

WHITE Peter — Transport Studies Group, Polytechnic of Central London, 35 Marylebone Road, London, NW1 5LS, UK.